STORMS
FROM THE
SUN

STORMS FROM THE SUN

THE EMERGING SCIENCE OF SPACE WEATHER

Michael J. Carlowicz
and Ramon E. Lopez

The Joseph Henry Press
Washington, DC

Joseph Henry Press • 2101 Constitution Avenue, N.W. • Washington, DC 20418

The Joseph Henry Press, an imprint of the National Academy Press, was created with the goal of making books on science, technology, and health more widely available to professionals and the public. Joseph Henry was one of the founders of the National Academy of Sciences and a leader in early American science.

Any opinions, findings, conclusions, or recommendations expressed in this volume are those of the authors and do not necessarily reflect the views of the National Academy of Sciences or its affiliated institutions.

Library of Congress Cataloging-in-Publication Data

Carlowicz, Michael J.
  Storms from the sun : the emerging science of space weather / Michael J. Carlowicz and Ramon E. Lopez.
     p. cm.
Includes bibliographical references and index.
  ISBN 0-309-07642-0 (hardcover)
  1. Space environment. 2. Sun—Environmental aspects. I. Lopez, Ramon E. II. Title.
  QB505 .C367 2002
  629.4'16—dc21
                                      2002003618

Cover photograph by PhotoDisc.

Copyright 2002 by the National Academy of Sciences. All rights reserved.

Printed in the United States of America

Lyrics from "Blinded by the Light" by Bruce Springsteen reprinted by permission. Copyright © 1972 by Bruce Springsteen (ASCAP).

*For Florence,
a shelter from this storm*

*For Nancy and Ramon III,
children of the space weather generation*

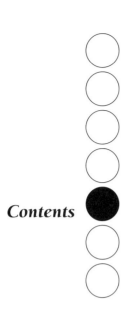
Contents

Foreword ix

Prologue: Here Comes the Sun 1

1 The Day the Pagers Died 11

2 Sun-Eating Dragons, Hairy Stars, and Bridges to Heaven 31

3 A Sudden Conflagration 51

4 Connecting Sun to Earth 61

5 Living in the Atmosphere of a Star 75

6 The Cosmic Wake-Up Call 93

7 Fire in the Sky 107

8 A Tough Place to Work 117

9	Houston, We Could Have a Problem	137
10	Seasons of the Sun	153
11	The Forecast	171
Epilogue: Over the Horizon		191
Appendix A: Selected Reading		199
Appendix B: Selected Web Sites		203
Appendix C: Acronyms and Abbreviations		207
Endnotes		211
Acknowledgments		217
Index		221

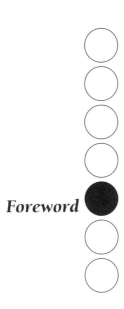

Foreword

Every day the rising and setting of the Sun regulate the lives of some 6 billion persons and a vastly greater number of other living creatures on the Earth. Indeed, "sunshine" is responsible for the long-term development and maintenance of the entire biosphere in all of its wondrous complexity and diversity. Not a bad performance for the quiescent thermal radiation of 1.36 kilowatts per square meter from a ball of gas at an effective temperature of 5,770 degrees Kelvin and a distance of 150 million kilometers!

And that is only the "first approximation," as scientists say, to the Sun's influence on the Earth.

The second approximation, now called space weather, encompasses a body of lesser phenomena most of which are not discernible to the unaided human senses and were therefore unknown to pre-technological peoples. The notable exceptions are the luminous visual displays in the Earth's upper atmosphere, termed the *aurora borealis* and the *aurora australis*.

This well-written and splendidly illustrated book by Michael J. Carlowicz and Ramon E. Lopez provides a timely and broadly intelligible overview of the nature of space weather. It is replete with historical background, anecdotal examples and citations of the recent contributions of many individuals. Above all, it is instructive for everyone who has an interest in the rapidly evolving science of space weather and its multifold implications in our technological culture.

The foundations of the subject have been under slow but progressive development for centuries but, until recently, have been of professional interest to only a small cadre of investigators who publish their findings in highly technical journals and specialized monographs. The 1957-1958 International Geophysical Year launched a major increase in worldwide scientific work in space physics. This increase has been sustained and accelerated in subsequent years.

In brief, space weather is attributable to the highly variable, outward flow of hot ionized gas (a weakly magnetized "plasma" at a temperature of about 100,000 degrees Kelvin, called the solar wind) from the Sun's upper atmosphere and to nonthermal, sporadic solar emissions of high-energy electrons and ions and electromagnetic waves in the X-ray and radio portions of the spectrum. There are also significant contributions by the relatively constant bombardment of the Earth by cosmic rays from distant sources in the galaxy. The power flow of each of these causes of space weather is far less than the previously quoted power flow of thermal radiation.

One of the early (1936) practical applications of space weather was the daily publication of maximum usable frequencies for short-wave radio communication by the U.S. National Bureau of Standards and by corresponding agencies in other countries. These advisories rested on routine observations of the partially ionized layers of the Earth's upper atmosphere, called the ionosphere.

The current and prospective importance of space weather in the realm of practical matters is attributable to the rapid advances that are being made in sophisticated applications of modern technology, both civil and military. Such applications include global telecommunications, large arrays of electrical power grids, navigation, reconnaissance and surveillance, forecasts of ordinary weather, safety of human space flight, and reliability and failure modes of hundreds of satellites of the Earth which serve these functions.

Despite voluminous scientific knowledge of the physics of magnetic storms and the sporadic emission of solar energetic particles, there is now a clear need for the application of this knowledge to informative and reliable forecasts. It is of central importance that forecasters work closely with engineers of vulnerable systems in order to reduce hyperbolic rhetoric and false alarms and to develop a mature and reliable body of forecasts that practitioners can take seriously. Improved knowledge of the fundamentals of space weather continues to be a largely governmental function, as is the case with ordinary weather, but it is now clear that increased commercial attention to alleviating the hazards of space weather is timely and appropriate. To this end, operators of electrical power grids and hundreds of Earth satellites and interplanetary spacecraft may well take this book to be a wake-up call for the design and development of invulnerability to adverse effects of space weather. Such engineering responses include identification and replacement of susceptible elements and the development of operational protocols, automatic self-checking and self-correcting circuits, and work-around software. Such measures are now made feasible by the very sophistication that has made modern space techniques of importance to our daily activities.

I commend this book to a wide spectrum of readers who may join me in enjoying a compelling account of the fascinating field of space weather and its evolving effects on our daily lives.

James A. Van Allen
University of Iowa
January 2002

STORMS
FROM THE
SUN

Prologue 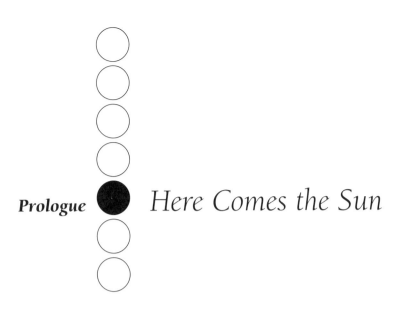 *Here Comes the Sun*

What has been will be again,
What has been done will be done again,
There is nothing new under the Sun.
                    Ecclesiastes 1:9.

In March and April 2001 the Sun went boom. Or maybe it was boom, boom, boom, boom, since the Sun unleashed a machine-gun burst of explosions and space storms. The surge of solar activity made the nightly news and newspapers, complete with speculation about northern light shows and communications problems. The outburst came just as the Sun was allegedly quieting down; the peak of solar activity, or "solar maximum," had officially arrived back in July 2000. The solar outburst in the spring of 2001 also coincided with some of the most extensive work ever performed on the International Space Station. The Sun hardly noticed man's solar calendar.

On March 22 a large sunspot rotated around the eastern edge of the Sun and into full view from Earth (see Figure 1). Scientists labeled the area "active region 9393," and for two weeks the group of sunspots roiled and pulsed. They swelled into a monstrosity that was visible to the naked eye at sunrise and sunset. The diameter of the active region spanned nearly 86,800 miles (140,000 kilometers), more than 22 times the diameter of Earth (see Figure 2). That same sunspot group lit up on April 2, produc-

FIGURE 1. The sunspot of March 2001 was so large that it was visible with the naked eye. Note that sunspots should only be observed through special filters; otherwise, you risk blindness, as many an ancient astronomer learned the hard way. A ground-based observatory in California collected this whitelight image. Courtesy of National Solar Observatory (NSO)/Association of Universities for Research in Astronomy (AURA)/National Science Foundation (NSF).

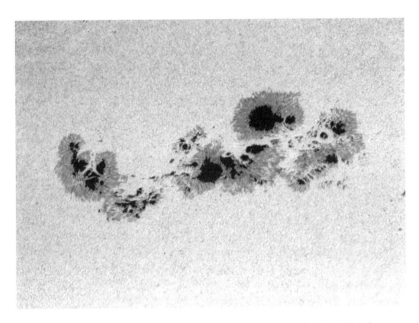

FIGURE 2. Solar active region 9393, shown here on March 30, 2001, became the largest sunspot group in a decade. The darkest structures are known as sunspot umbrae, where the magnetic fields point out vertically from the Sun; the lighter fibrous channels are called penumbrae, where the magnetic fields are horizontal to the surface. The surrounding bubble-like structures are known as photospheric granules, each of which is about 620–1,240 miles (1,000–2,000 km) across. Courtesy of NSO/AURA/NSF.

ing a rare "white-light" flare that was visible from Earth. The explosion was the most intense flare observed since scientists first began to keep records of the X-ray intensity in 1976. Scientists rated the flare an X-22 event on a scale that's only supposed to go to X-20.

By the time region 9393 started rotating around the western limb to the back side of the Sun, another hot spot rolled into view on the eastern edge. In the first week of April, active region 9415 burst onto the scene with its own monster sunspot and its own flurry of storms. Two weeks later, region 9393 rolled back around to the front of the Sun, now named active region 9433, and bearing most of the same ugly gnarled sunspots that first appeared in

March. All told, the Sun stormed for nearly six weeks, blazing with dozens of solar flares and spitting several blobs of superheated gas, known as coronal mass ejections, or CMEs, toward Earth.

As a direct result of the storms on the Sun, a storm raged in the space around Earth and northern lights danced as far south as El Paso, Texas, and Southern California. Radio communications were distorted and occasionally blacked out over parts of the world for operators using high-frequency radio signals, such as airlines, ship-to-shore radio, and the Voice of America and BBC World Service. At least two U.S. military satellites and several commercial satellites suffered outages, hardware failures, and computer errors. Some electric power companies worked to reroute power supplies so that their equipment would not be overwhelmed by surges of electric power from space, and several transformers tripped in New York and Nova Scotia. Commercial airlines had sporadic problems with their radio signals during flights over the Pacific, and more than 25 flights between North America and Asia were diverted so as not to fly through the polar regions, where radio communication is more susceptible to blackouts and passengers are more exposed to solar radiation. The National Aeronautics and Space Administration (NASA) even considered delaying the launch of its 2001 Mars Odyssey spacecraft, which was scheduled for an April 7 liftoff. With the Sun spewing unusually high doses of radiation—particularly high-energy particles that can wreak havoc with the electronics on satellites and rocket boosters—mission controllers did not want to risk a faulty signal or computer error that could have foiled yet another Mars mission. The Sun eventually cooled off enough for the launch to take place, but not without a few days of contingency planning and heartburn.

In the midst of all of this solar storming, NASA and Russia were welcoming the second astronaut crew to International Space Station *Alpha* and preparing to launch another space shuttle mission to continue construction. The crew of Expedition Two—Yury Usachev, James Voss, and Susan Helms—arrived at the station on March 9, and two weeks later, active region 9393 began storming. When active region 9415 took over the fireworks, it launched an intense solar flare and coronal mass ejection on April 18, bathing the Earth in solar particles. The next day the space shuttle *Endeavor* lifted off for a rendezvous with the space station and a major construction project. Space shuttle mission STS-100 included two space walks. For a little more than 26 hours, the astronauts roamed outside the station and installed the Canadarm2 robotic arm and the Raffaello logistics module, a sort of space-age moving van. NASA called the mission the most intricate and advanced robotic installation and operation ever conducted in space. It also coincided with the second most active period of solar activity in the entire solar cycle.

All through the six weeks of solar storms, physicists at NASA's Johnson Space Center kept a watchful eye on the human occupants of the space station and the shuttles. Scientists monitor the radiation doses that crews receive on every space flight, and the March and April missions demanded acute attention. NASA's Space Radiation Analysis Group monitored the Sun and the space around Earth with extra care during the April shuttle flight because of the extravehicular activity (EVA or, more simply, a space walk). When the Sun spits a flare into space toward the Earth, the X rays and accelerated radioactive particles can arrive at Earth in no less than 10 minutes and no more than 30. That's not a lot of time to get an astronaut into a shuttle when he or she has been walking in space. High-energy protons from a solar flare can pierce a space suit with little trouble, causing damage to human cells and tissues that could be minor or deadly.

"With current planning and flight rules, we believe the current dose of radiation to the crew inside the space station is too small to be of concern," said Dr. Gautam Badhwar, NASA's chief scientist for space radiation at the time. "There is a definite but very small increase in the probability of cancer induction and mortality due to the exposure. But serious attempts are made to minimize these by flight rules." In other words, if the Sun is spewing too much harmful radiation, flight controllers can postpone space walks and keep the astronauts hunkered down in the deepest, most shielded parts of the spacecraft. But in the spring of 2001, that was not deemed necessary because the most severe storms never coincided with activities outside the shuttle or space station.

"The one possibility for radiation sickness might be an EVA situation during a solar event, if perhaps a crew member couldn't be brought back inside safely," Badhwar added. In such an event, controllers would direct space-walking astronauts to hurry back inside before the solar storm arrives, but that could take more time than nature allows. No one knows for sure what would happen to an astronaut caught outside the spacecraft during a storm (because it has never happened), but the best guesses range from skin burns and cataracts to Hiroshima-like radiation sickness.

Through good luck and watchful monitoring, the astronauts were able to safely conduct their space walks as planned in April. After the flare of April 18, the Sun was relatively quiet for more than a week, allowing the space station construction crew to complete their work without delay or danger. Ironically, the crew from *Endeavor* delivered several science experiments to the space station designed to assess some of the problems that will be encountered on longer future missions to the Moon or Mars. Germany's DOSMAP dosimetric mapping experiment was used to delineate the types and intensities of radiation that seep into the space station, while Japan's Bonner Ball neutron detector measured the number of uncharged atomic particles reaching the inside of the station. Energetic neutrons can penetrate human tissues and can affect the development of bone marrow, leading to blood diseases and cancers.

Gautam Badhwar had an experiment included in that care package from Earth. The "Phantom Torso" experiment was a mock-up of a human body that was similar to the dummies used by radiologists as they are trained to use X rays on Earth. The phantom—named Fred—was designed to replicate the head and upper-body anatomy of an adult male. The 3-foot, 95-pound model was composed of artificial human bones and skin, and organs—brain, thyroid, heart, lungs, stomach, and colon—made to closely match the density of real organs. Each of the 35 sections of the torso and each organ included radiation detectors, allowing

Badhwar and colleagues to monitor radiation doses in real time during the flight and long after the experiment was over. Phantom Fred was the first experiment to measure doses of radiation and the effects *inside* the human body, particularly the blood-forming organs. In the past, human radiation experiments in space (on *Mir* and on shuttle flights) have observed the amount of radiation that reaches the skin and then extrapolated those numbers to project the amount that reached the organs. The hope is that scientists will finally get a somewhat realistic picture of how much space radiation seeps into the body during space flights, a picture that heretofore could only be guessed at.

Two hundred years ago, before telegraphs and telephones, before rockets and radio, very few people were affected by space weather. As journalist John Brooks wrote in *The New Yorker* after a famous storm in 1958: "Down through the ages, [magnetic storms] no doubt raged over the Earth at intervals, confusing mariners by causing compass needles to waver, but otherwise spending their force on a planet too electromagnetically innocent to suffer from them." Solar activity and the resulting changes in the environment of Earth were scientific curiosities, engaging mysteries for the men and women who observed the stars, navigated ships, or studied magnetism.

But then we started to harness the power of electromagnetism. We developed electric power and communications systems and built worldwide networks that are now critical for our civilization but also vulnerable to space weather. We put ourselves in harm's way, yet very few people even know there is a problem.

In order to understand and appreciate space weather, we need to grasp three basic concepts that are mostly left out of our science books and classes. First, the Sun is a dynamic, variable star. It may seem like old reliable, barely changing from sunrise to sunset, from year to year. But when we look at it closely, with the tools of science, we see that the Sun is constantly changing, with patterns ranging from seconds to centuries. In addition, Earth resides within the atmosphere of that Sun. Stretching well beyond Pluto and the outermost rocks and ice balls of our solar system, the atmosphere of the Sun swirls around all of the planets, carrying the imprint of its energy and magnetism in a solar wind and in the periodic explosions such as flares and coronal mass ejections. Finally, Earth responds to this changing Sun and its turbulent atmosphere. Activity on the Sun shapes, distorts, and influences the environment around the Earth. Our planet is a giant magnet, and that invisible magnetic force (the force that makes a compass needle point north) shelters us from many of the Sun's most harmful effects. But once we start working in space, when we launch satellites and start walking in space, we move closer to the source of space weather, closer to the eye of the storm from the Sun.

Space weather is a natural hazard, one that matters to anyone who works in space or uses space to work. By using technology based on electromagnetism to communicate, navigate, predict the weather, study our environment, and defend our nations, we also have created new vulnerabilities. Every benefit must have its cost, so every tool or gadget that relies on radio waves, conducting wires, and sensitive transistors and microchips can be affected by electromagnetic disturbances in the Sun-Earth system. Space weather distorts radio and television signals as they bounce around the atmosphere. Magnetic storms can add unwanted electricity into our power lines and pipelines, causing blackouts and brownouts and fuel leaks. And because these storms can damage spacecraft, things that we now take for granted—like the Global Positioning System or our worldwide communications systems—can suddenly stop working.

An environment of which most of us are not even aware affects our economies and daily lives. We are increasingly dependent on space-based infrastructure for the humdrum things of daily life. When we watch a sitcom or sporting event on TV, when we wait by the phone for the pediatrician to answer her page, when we navigate a ship through a narrow channel or land an airplane in low visibility, when we flash a speed-payment pass at the gasoline pump, more likely than not we are relying on space-based technology. Yet space is a place where in the blink of an eye a maelstrom can erupt, with potentially severe consequences for those technologies. Our modern electronic society is fragile, as we discovered several times in the 1990s. We are exposed and vulnerable to the whims of a star that so far defies prediction.

So it is to the citizen of the twenty-first century to whom this book is addressed. We are now a space-faring race, and we have moved beyond simple voyages of discovery. We have gone to work in space, and in time the notion of space weather will become commonplace, perhaps a feature of the nightly weather reports. We are living with a star, and if you like your electronic toys and tools—or if you work for or invest in the companies that make them—you ought to learn something more about your nearest star. The Sun is the only star we can study up close, and it is probably the only one that will affect you in your lifetime.

1   The Day the Pagers Died

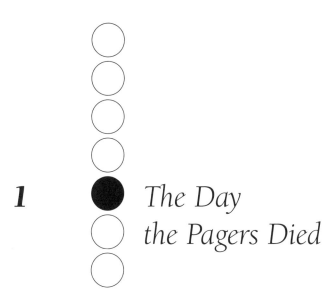

> Let me leap out of the frying pan into the fire;
> or, out of God's blessing into the warm sun.
>   Miguel de Cervantes, *Don Quixote*

On May 19, 1998, modern, high-tech society was zapped back into the phone age. Shortly after 6 p.m. on the East Coast of the United States—while evening commuters were tuning their radios to National Public Radio's (NPR) "All Things Considered," while police, rescue, and fire crews were easing into the second shift, while hospitals and obstetricians' offices were paging their doctors for emergency surgeries and unexpected deliveries—a satellite died. Within moments, NPR and CBS News were scrambling to fill dead air; within hours, thousands of doctors, detectives, and drug dealers were discovering that they had been out of touch all night. By the next morning, newspapers and morning talk shows were buzzing about the day the pagers died.

For reasons that the owners of the Galaxy IV communications satellite could not explain, the $200 million messenger for nearly 90 percent of North America's pagers and several major broadcast networks had been reduced to space junk. Without a satellite to relay radio signals from dish to dish, paging companies could not deliver messages to their customers. Radio and television stations could not deliver their signals to their affiliates. Millions of North Americans were suddenly forced to live without some of their most prized gadgets. Never before had so many people paid attention to the life and death of a satellite floating 22,000 miles (35,000 kilometers) overhead.

A pair of unrelated incidents, as the PanAmSat Corporation would announce months later, had destroyed the crucial computer processors that kept the satellite pointing at Earth. One computer failed due to the growth of crystals in its circuits. The other failed due to a "random event" that the company could not explain. And since the satellite could not be plucked out of space and examined, no one would ever really know. So without a means to point and orient itself, the satellite went into a spin, lost its way, and became a spare part drifting in space.

To the scientists who study the space around Earth—also known as geospace—the failure of Galaxy IV was no random event. Gathering evidence from half a dozen government research satellites, space scientists proposed their own cause of death for Galaxy IV. In the weeks leading up to the failure of the pager satellite, a series of explosions on the Sun severely disturbed the space around Earth. The barrage left the uppermost reaches of Earth's atmosphere teeming with "killer" electrons—the kind that murder satellites.

○●○

From April 27 to May 6, 1998, the face of the Sun lit up with the solar system's most common sort of fireworks display. According to Barbara Thompson, a space weather researcher at NASA's Goddard Space Flight Center in Greenbelt, Maryland, the Sun spat super-hot gas and radiation into space more than 90 times in the two-week span, much of it directed at Earth. And in the days following those solar eruptions, the space around Earth was buffeted by a series of intense space weather storms.

Hardly a static, benevolent body, the Sun constantly seethes with activity, producing the cosmic equivalent of hurricanes and tornadoes. The most famous of these weather events is the solar flare, which scientists and ham radio operators have long blamed for radio disturbances. These bright flashes on the face of the Sun are explosions that release the force of 10 million volcanic eruptions in a matter of minutes. Provoked by the buildup of magnetic energy in the Sun's atmosphere, flares spew radiation—including radio waves and X rays—and propel some superheated particles across space.

More important from the vantage of Earth, Thompson notes, is the coronal mass ejection (CME), the solar equivalent of a hurricane. Once a graduate student who focused on the physics of Earth's magnetic field (the magnetosphere), Thompson now looks upwind from our planet, trying to understand how and when CMEs might impact Earth. A CME, she explains, is the eruption of a huge bubble of plasma from the Sun's outer atmosphere (the corona). CMEs are the principal way that the Sun ejects material and energy into the solar system, and they are the largest structures that erupt from the Sun.

Most CMEs travel at 1 million miles per hour, but the fastest bursts can reach a speed of 5 million miles per hour. But it's not the speed that kills. A typical CME carries more than 10 billion tons of hot, electrically charged gas into the solar system, a mass equal to that of 100,000 battleships. And the amount of energy trapped in just one of these bubbles is greater than the energy of 1 billion megatons of TNT. Having observed CMEs since 1996 through the telescopes on the Solar and Heliospheric Observatory (SOHO) satellite, Thompson and her colleagues know that these bubbles of solar plasma can pack a punch comparable to that of 100,000 hurricanes.

Just hours after blowing into space, a CME cloud can grow as wide as 30 million miles across, 35 times the diameter of the Sun itself. As it speeds away from the Sun, the CME drives like a plow into the slower, steady solar wind of electrically charged particles, piling the wind in front of it like snow on the blade of the plow. This plow-like effect creates shock waves that accelerate some atomic particles drifting in space to dangerously high energies. Right behind that potent shock, the CME cloud flies through the solar system bombarding planets, asteroids, and other objects with radiation and plasma. If a CME travels on a path that intersects the Earth's orbit around the Sun, the results can be spectacular and sometimes hazardous.

Conditions were right for such hazards in the spring of 1998. Around April 27, a roiling, turbulent region (officially named active region 8210) popped up near the equator of the Sun, complementing two other active regions on the visible edges of the Sun (see Figure 3). In the next two weeks, instruments on the U.S. Geostationary Operational Environment Satellites (known as GOES, these spacecraft capture the satellite weather images shown on TV) detected 86 flares exploding from the side of the Sun facing Earth. Thompson observed that nearly three-quarters of the flares originated in that one active region—and eight CMEs were spit out of that same hot spot.

But all of the fireworks might have been just a pretty spectacle had they shot off from a different location. In early May, Thompson noted: "Active region 8210 was magnetically well connected to Earth." As the Sun spews solar wind and its various ejections into space, the material tends to fly off in spirals that are curved by the rotation of the Sun. Earth was sitting at the far end of the spiral (a magnetic field line) that began at 8210, putting it on a celestial beeline from the Sun.

"What was unusual in May was that the active region on the Sun was directly facing the Earth when it let off a bunch of CMEs," says Thompson, whose past work with space physics has helped her bridge the gap between researchers who focus on only the Sun or on only Earth's magnetosphere. "There were a lot of energetic

The Day the Pagers Died

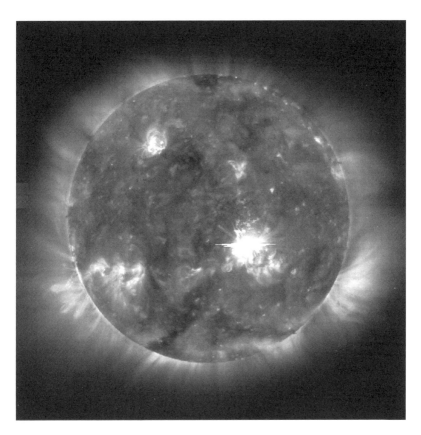

FIGURE 3. The Sun fired dozens of flares and coronal mass ejections into space in April and May 1998, sending a warning shot that solar maximum was on the way. This SOHO image of the Sun in extreme ultraviolet light depicts the flare of May 2, the most potent of the storm. Courtesy of SOHO/European Space Agency and NASA.

CMEs coming from the same region. We see a lot of these kinds of events as magnetic flux emerges on the Sun and forms complex regions during solar maximum."

One of the first victims of active region 8210 was the Large-angle Spectrometric Coronagraph instrument on SOHO. The camera was flooded with radiation—in the form of protons accelerated by the solar snowplow—several times over a two-week span.

The spacecraft's image-processing computers slowed to a crawl because there was just too much to see on the Sun. It would take several hours to get them working again.

Downwind from SOHO, NASA's Wind spacecraft recorded solar wind blowing faster than at any time since the spacecraft had been launched in 1994. Between May 1 and May 19, Wind recorded four separate solar wind streams exceeding 1.3 million miles per hour (600 kilometers per second). On May 4 the solar wind speed peaked at nearly 2 million mph (850 km/s).

On May 4 all of the indexes that measure magnetic and electrical activity in the space around Earth went wild. The auroral electrojet index, which measures the strength of the magnetic disturbances produced by currents in the ionosphere (associated with auroras), reached peak values of more than 2,000 nanoTesla. In the first six hours of the day about 4,600 gigawatt-hours of electrical energy were dumped into the upper atmosphere—more than the total electrical energy produced by all U.S. power stations combined during the same six hours. Another 3,000 gigawatt-hours were dissipated in space to create a huge "ring current" around the Earth. These disturbances had visible effects. On May 4 space weather brought auroras to Boston, London, and Chicago and forced electric companies to reconfigure their power grids in New England.

As detected by several science satellites, the leading edge of Earth's magnetic field (magnetosphere) was pushed down to 15,300 miles; it normally stretches about 45,000 miles from Earth toward the Sun. Some geosynchronous satellites—which orbit the Earth at 22,300 miles altitude—well inside the magnetosphere and protected from most of the harshest elements of Sun and space—were periodically left twisting in the solar wind. The satellites didn't move; their whole neighborhood in space moved away for a while.

On the ground in frigid Halley Bay, Antarctica, the storms from the Sun produced their own intrigue. "The *aurora australis* was flickering all over the sky," says Matthew Paley of the British Antarctic Survey, a scientist and aurora watcher. "I could read the label on the back of my gloves by aurora light."

"We usually get 10 to 20 nights of good aurora in a typical winter, plus frequent steady arcs over the southern horizon," Paley notes, arcs that are miles away, that result from moderate storms. "The first time people see one of these steady arcs it is an automatic waste of a roll of film." Such distant arcs are almost boring to the Antarctic veterans, who know that the dancing curtains will eventually appear straight overhead during an intense space weather storm. "The significant auroras are when the arc expands, brightens, and breaks up into complicated waves with bright rays climbing up into the sky," he added. "These auroras can move over the entire sky and back again in a few seconds, with flickering and bright blobs moving from horizon to horizon."

More commonly known as the northern and southern lights—the *aurora borealis* and *aurora australis,* respectively—these light shows are relatively benign and beautiful signs of Earth's electric relationship to the Sun. Named for the Roman goddess of dawn, these dancing ribbons and rays of light appear most often in skies above the northern and southernmost parts of the world—Canada, Alaska, Scandinavia, New Zealand—between 65 and 75 degrees of magnetic latitude (from most of Antarctica, that means you have to look north). Auroras are provoked by energy from the Sun and fueled by particles trapped in Earth's magnetosphere. Like sunspots, they are the space equivalent of the canary in the mineshaft, warning of unseen trouble in the regions around Earth. When the Sun and the space around Earth are really active, auroras have been observed to appear as close to the equator as Singapore.

Auroras are formed by a process that is similar to the way fluorescent lights and televisions work. Some of the currents that flow through the magnetosphere can flow down along the Earth's magnetic field to form two ovals of current centered on the north and south magnetic poles. If the currents are large enough—such as when they are pumped up by a CME—they cause electric fields to be set up along the magnetic field. The electric fields in turn accelerate particles, which plunge into the upper atmosphere (an electrified region known as the ionosphere) where they collide with oxygen and nitrogen molecules and atoms. These collisions—which usually occur between 60 and 200 miles above ground—cause the oxygen and nitrogen to become excited and emit light.

The result is a dazzling dance of green, blue, white, and red light caused by the different elements in the atmosphere. Auroras can appear as colorful, wispy curtains of light ruffling in the night sky or sometimes as diffuse, flickering bands. Sometimes they make rays and arcs. In any form, they tell us that something electric is happening in the space around Earth.

"We have a system of wake-up calls where the night watchman wakes the keener people up whenever there is a good aurora," Paley says. "We generally grab cameras, put on loads of clothes, and go outside to lie on our backs in the snow, which by popular consent is the best position to truly appreciate the spectacle. After 10 to 20 minutes of lying in the snow at 35 degrees below zero, it just gets too cold and people start going inside to defrost before coming out again until they get too cold, tired, or the aurora fizzles out."

May 4, 1998, was one of those nights when the watchman called. Paley's journal from that night includes more than eight hours of aurora observations. "The May 4 event looked especially impressive because the arcs were exceptionally bright, numerous, and changeable," he noted. "The sky also was lit by a diffuse background aurora. When the main arcs surged to a peak in brightness, small details such as the writing on my gloves and my footprints in the snow were clearly visible in the green glow. It is quite common to be able to see silhouettes of people and snowmobiles. But to be able to see clearly—as if by bright moonlight—is really unusual."

The *aurora australis* that Paley witnessed that night coincided with one of the larger natural changes of the Earth's magnetic field ever recorded from Antarctica. The flow of electric currents in the ionosphere decreased and offset the strength of Earth's magnetic field (as detected on Earth's surface) by more than 10 percent. From May 2 to 6, high-frequency radio communications were impossible in the Antarctic because of all of the mayhem in the ionosphere. But at least they had auroral light to read by.

As Dan Baker puts it, blowing a CME across the bow of Earth's magnetosphere is like exploding a nuclear weapon in space. More specifically, it's like exploding a bomb in the Van Allen radiation belts, the vast rings of electrically charged particles that swirl in the space near Earth.

For more than 25 years, Baker has been interested in the energy and atoms trapped in the space around Earth. From the radiation belts to the ionosphere to the distant tail of Earth's magnetosphere, a certain amount of high-energy particles are always suspended above our atmosphere. As an investigator for NASA's Solar, Anomalous, and Magnetospheric Particle Explorer (SAMPEX) and for the International Solar-Terrestrial Physics program—and as a product of the University of Iowa, where Space Age pioneer James Van Allen held court—Baker has come to know much about the radiation belts.

Discovered in 1958 by Van Allen, the radiation belts are a pair of doughnut-shaped rings of very energetic particles and hot ionized gas (plasma) trapped in the external magnetic field of Earth. Magnetic fields exert forces on electrically charged objects as they move. For charged particles (protons and electrons) that approach near-Earth space, the magnetic force is much stronger than the force of gravity. So when particles arrive from the Sun, Earth's magnetism takes control of their motion. Earth's magnetic field is very much like the magnetic field of a bar magnet, where the field comes out of one pole and goes into the other. After charged particles arrive from the Sun and get caught up in Earth's magnetic field, they spiral along these lines of magnetic force, bouncing from north to south and back again. At the same time, the magnetic field forces the particles to drift roughly in a circle around the Earth (protons drift to the west, electrons to the east). And so the rapidly moving particles are effectively trapped in doughnut-shaped regions around the Earth, creating the belts of radiation centered around the magnetic equator of the planet.

According to Van Allen, the outer radiation belt stretches from roughly 7,000 to 40,000 miles above Earth; the inner belt lies lower in Earth's space, inside the outer belt, between 300 and 7,000 miles above Earth's surface. In the higher latitudes of Earth, near the Arctic and Antarctic Circles, the belts connect to the uppermost portion of the atmosphere (the ionosphere), where they dump those particles that stray too far down the magnetic field lines.

Since the radiation in the belts poses a potential threat to astronauts and to sensitive electronics, engineers have tried to design satellites and human space missions that could overcome the hazards of working in space. Lacking consistent real-time information about what was happening in the belts, they built models and made predictions of "average" intensity. Yet scientists and engineers have known for decades that while the inner belt is relatively stable, the outer belt can vary on relatively short time scales—so conditions are rarely "average." The best solution was to build "hardened" satellites and spacecraft that could endure a steady dose of radiation (an expensive endeavor), or to place spacecraft in orbits that would avoid spending much time in the high-intensity regions of the Van Allen belts. For instance, satellites in geosynchronous orbit, such as Galaxy IV and most of our weather and communications satellites, reside on the outer edge of the outer radiation belt, where conditions are tolerable most of the time. International Space Station *Alpha* and the space shuttle are both flown well beneath the inner belt, though they occasionally cross the regions where the belts snuggle close to Earth, such as the South Atlantic magnetic anomaly.

In more comprehensive studies since the last peak of the solar cycle in 1989, Dan Baker and colleague Geoff Reeves have found that the intensity of the belts can actually vary by 10, 100, or even 1,000 times in a matter of seconds to minutes. And regions thought to be safe can, on occasion, be deadly to satellites. "The radiation belts are almost never in equilibrium," says Reeves, a space physicist at Los Alamos National Laboratory who spends much of his research time examining space radiation. "We don't really understand the process, but we do know that things are changing constantly."

"It's amazing that the system can take the chaotic energy of the solar wind and utilize it so quickly and coherently," says Baker, head of the Laboratory for Atmospheric and Space Physics at the University of Colorado. "We had thought the radiation belts were a slow, lumbering feature of Earth, but in fact they can turn on a knife's edge." No one has offered a sufficient explanation for the dramatic shifts that the belts can make.

On May 4, 1998, the Van Allen belts took one of those sharp turns, and they did not get back on track for six weeks. In the wake of Barbara Thompson's solar blasts and Matt Paley's dancing auroras, the relatively harmless atomic particles that are naturally suspended in the belts were suddenly excited to high energies. According to measurements from spacecraft and ground observatories, a magnetic storm was just beginning to wane when a tremendous shock wave from the Sun arrived on May 4. After being hammered by solar shocks and clouds for a week, then punched one more time, the electrons of the outer radiation belt were screaming around the Earth with energies in the millions of electron volts and at velocities approaching the speed of light. It was as if the electrons and ions had been run through a laboratory particle accelerator.

Driven by processes that are still not understood, the energetic electrons overflowed the two usual radiation belts and also became trapped in between. By the end of May 4, data from NASA's SAMPEX and Polar spacecraft and several Defense Department satellites showed that a new radiation belt had formed around the Earth. The new belt would last until mid-June.

"We may not be living on the Moon, but society is expanding into space," says Geoff Reeves, who is developing what he calls "weather maps" of the radiation belts. "Your wireless phone may be routed through a satellite that has to operate in the radiation belts. The Earth's radiation belts are in turn connected to things that happen on the Sun."

"We are going to work in space, to borrow an old space shuttle slogan," says Reeves. "And much of that work happens on the fringes of the radiation belts." Knowing something about how the radiation belts behave, and knowing how wild they were in May 1998, Baker and Reeves and other colleagues believe they can make a compelling case that Mother Nature—not some mechanical or engineering error—probably crippled the world's pager network. They point to a lengthy list of timely coincidences—Baker sees it as circumstantial evidence—to suggest that the cause of Galaxy IV was something less than mysterious and "random."

While the Sun was raging in April and the radiation belts were tightening in May, one major science satellite failed (Germany's Equator-S) and several others suffered blackouts and operational problems (NASA's Polar and Japan's GMS satellites). News accounts from trade newspapers indicate that Motorola lost four of its Iridium satellites in the second quarter of 1998 (April to June), before they were even put online, and the company could not determine a cause for two of the losses. Galaxy IV failed in the midst of all this activity. And according to one dean of the space physics community, Chris Russell of the University of California at Los Angeles, "the guilt by association is really strong."

Getting subatomic particles to take out minivan-sized satellites is complicated but not uncommon when the Sun is active, Baker notes. The process is quite similar to shocking or sparking a friend on Earth. Like the 10-year-old dragging his feet on a carpet, a satellite flying through electron-filled space collects electric charge on its surfaces. That charge can build up to a point where it is released quickly as a damaging spark. If sensitive electronic parts or cables are exposed, the shock can upset the spacecraft. Well aware of this problem, engineers design spacecraft with devices or orbit maneuvers that reduce the amount of charge on the surface and keep it from affecting the parts inside.

High-energy electrons, like those swirling in the radiation belts in May 1998, can cause more insidious problems. Baker and colleagues call the process deep dielectric charging. Dielectric materials—insulators that prevent the flow of direct electrical current—cover many of the wires and electronics inside satellites. As satellites drift through space, the ever-present low-energy electrons tend to accumulate and stick to the outer surfaces, never reaching the sensitive electronics inside. High-energy electrons—a reporter once dubbed them "killer electrons" and scientists kept the name—can penetrate the shell of the spacecraft and lodge themselves in the dielectric insulation around the electronic nerve centers (computer chips, coaxial cables, etc.). Under normal circumstances, very few high-energy electrons lodge in the dielectric materials, and most naturally leak away over time. But when high-energy electrons overwhelm a spacecraft, the sparks can fly, frying million-dollar equipment.

"Dielectric charging can cause, or at least exacerbate, problems with spacecraft electronics," asserts Baker, who is accused by some industry colleagues of howling at the space weather Moon. He has closely studied several spacecraft failures, such as the demise of Canada's Anik E1 and E2 satellites in 1994, and has found strong evidence that long periods of exposure to concentrated radiation can wreak havoc on satellite hardware. Even hardened U.S. Air Force satellites have suffered from long-term, high-energy particle baths. "No matter if there is some other flaw in the satellite," Baker notes, "charging can push a touchy system over the edge."

In an article published in the scientific newspaper *Eos* in October 1998,[1] Baker, Reeves, and Joe Allen (former head of the solar-terrestrial physics division of the National Geophysical Data Center, part of the National Oceanic and Atmospheric Administration) dared to suggest what most space weather researchers usually only claim in private. In print for all of their colleagues, they made the case that Galaxy IV was likely overwhelmed by more killer electrons than it was designed to handle. Anticipating critics who would claim that there was no single day when the amount of killer electrons reached a precipitous peak, the scientists observed that the May event was one of the longest and most extreme cases of interaction between Sun and Earth in the 1990s. The amount of high-energy particles in the magnetosphere was well beyond the norm for more than three weeks. The result, the scientists asserted, was a slow, wasting death for Galaxy IV.

Geoff Reeves explained it this way. "If you punch some holes in a coffee can and leave it out in a slow rain or drizzle, the can will never get full," he says. "But leave it out in a downpour and very soon the can will be overflowing." He believes Galaxy IV was probably drowned in the steady downpour of killer electrons.

Without access to proprietary data from PanAmSat or Hughes Space and Communications (which built Galaxy IV and 40 other satellites of the same class), space scientists and most of the public will never know for sure what killed Galaxy IV, a satellite that was expected to last another five to 10 years. But after speaking with colleagues in the satellite industry, many in the science community remained confident that the dielectric-charging hypothesis was sound. In fact, the head of the National Oceanic and Atmospheric Administration, James Baker, made a point of highlighting the failure of Galaxy IV during a November 1999 press conference even as colleagues from his industry sat next to him.

"None of these things can be proven definitively," says Gordon Rostoker, a colleague of Baker and Reeves who runs a network of magnetic observatories in northern Canada. "After all, the satellite is 33,000 kilometers up there, and you can't go out and check the circuits. But when you have all of these birds having all of these problems, to say it's just engineering is silly."

As private companies, PanAmSat and Hughes do not have to release reports from their internal investigations. And as a player in the multibillion-dollar satellite communications industry, Hughes has an interest in attributing the failure to poor engineering and random events. Engineering failures can be covered by insurance policies and prevented in future satellite designs. Acts of God or the Sun are not always insurable or preventable.

○●○

In the *Eos* article, Baker, Reeves, and colleagues noted that, "whether or not the incident on May 19 was caused by space weather, it nonetheless shows the vulnerability of society to individual spacecraft loss. The vast number of users affected by the loss of just one spacecraft shows how dependent society is on space technology and how fragile communications systems can be." About 80 to 90 percent of all North American pagers were affected by the failure of Galaxy IV, according to a spokesman for PageNet, one of the nation's largest paging companies. Between 45 million and 49 million people use pagers, and at least 40 million of them lost that privilege on May 19. Doctors and nurses could not be reached for emergency calls. Real estate agents and corporate executives missed important messages. Drug pushers missed a few deals. It was the first major outage in 35 years for the paging industry.

The effects of Galaxy IV's early demise spread well beyond the pager networks. NPR depended wholly on the satellite to distribute its popular "All Things Considered" news program to stations across the country. At the time of the outage, the two-hour show was in the midst of its second broadcast feed of the afternoon, with one more to go before the afternoon drive time would be over. When Galaxy IV shut down, NPR stations endured several minutes of dead silence before switching to alternate satellites, phone lines, and the Internet to get the program out to local stations.

CBS radio and television, the Chinese Television Network, and CNN's Airport Network all lost signals routed through Galaxy IV. Reuters news service lost some of the reports it was sending through the communications satellite. And the satellite carried important weather information to numerous industries and agencies. Some radar signals and imagery for aircraft, agriculture, commodities traders, farmers, and emergency managers were temporarily wiped out for large portions of the United States. At airports, many airlines (including American, United, USAir,

Continental, Delta, Federal Express) lost some of the weather-tracking information they use to plan routes and forecast arrival times. Some air traffic control data were knocked out, as well, although the Federal Aviation Administration assured everyone that the data were not critical.

How could such a large communications network be brought down by one satellite? Stationed in prime space real estate at longitude 99 degrees W (above Kansas), Galaxy IV was in the perfect position to reach the entire United States with almost no interference from mountains, buildings, or even the horizon. And because of its signal capacity, the satellite was able to link up many clients, particularly pager networks because they use so little bandwidth. PanAmSat had pitched Galaxy IV as the ultimate satellite for paging services, and most companies bought the idea—so many that an entire industry collapsed, albeit briefly.

In the days and weeks after the Galaxy IV failure, PanAmSat initiated a contingency plan by moving the dead satellite to a higher orbit and moving another of its satellites, Galaxy VI, to Galaxy IV's old neighborhood. Within days, most of PanAmSat's customers were back online.

Meanwhile, engineers from PanAmSat and Hughes Space and Communications conducted an "extensive analysis of the cause of the spacecraft control processor failures." In the 12 years since Hughes had begun building and flying the HS 601—the model of satellite that included Galaxy IV—there had never been an "operational failure." In fact, since 1963 no other Hughes-built satellite of any kind had failed.

Three months after the May space weather barrage, officials from the two companies announced that the demise of the communications satellite was caused by two separate problems. Laboratory tests of spacecraft components suggested that Galaxy IV's primary control system was felled by the formation of crystals on the system's tin-plated relay switches. Barely the width of a human hair, the "tin whiskers" had caused catastrophic electrical shorts in Galaxy IV's circuits. Identical shorts in tin-plated relays would later disrupt PanAmSat's Galaxy VII satellite in June 1998 and one of Hughes's DirecTV satellites in July 1998, though in each case the satellites successfully switched to backup systems.

In May 1998, Galaxy IV had no backup ready to take over when the primary system failed. Analysis of components and data from an identical test satellite in the lab was inconclusive. So in an August 1998 announcement and in financial reports filed by PanAmSat, the company declared the failure of the backup to be a "different, isolated incident" from the tin-plated switch problem. Company officials called it a "one-time, random event." Some satellite engineers privately asserted that the second failure was also a "tin-whiskers" problem. But as Joe Allen points out, the failure need not be pinned solely to engineering or the environment. "If the solder joints and sockets of circuit board components develop these extended whiskers under the temperature and pressure conditions in orbit," Allen notes, "and they create a narrow gap between an active circuit and another conductor, across which a spark can jump with greater facility than would otherwise happen—then wouldn't you expect to see both tin whiskers and disturbed space environment conditions happening at the same time? This is likely a case where the environment exacerbated an engineering problem. Usually the satellite builders like to emphasize engineering, and I emphasize the space environment, and we're both right to some extent."

Though PanAmSat lost tens of millions of dollars in business and in hardware costs, the satellite itself was insured for $165 million, lessening the blow somewhat. But the damage does not necessarily end with the replacement of the satellite or the redirection of business. Investors grow wary of companies with satellites that are perceived to be vulnerable—even if the failures are rare. Competitors remind potential customers that "our satellites don't fail," even if they just happen to be lucky to have escaped the latest storm.[2]

During the past decade, the failure rate of satellites has been about 1 percent, according to William Kennard, former chairman of the Federal Communications Commission. "This type of disruption in satellite service is extremely rare." But Kennard's assessment fails to take into account that most of the current commercial satellites were launched during the relatively mild middle of the 11-year solar cycle.

With solar activity reaching a crescendo in the solar maximum period of 2000 to 2002, satellites are under more stress than they ever endured during the comparatively benign solar minimum of the mid-1990s. And there will soon be a lot more satellites to worry about, with 1,200 expected to be in orbit by the middle of the first decade of the twenty-first century (compared with 650 at the end of the twentieth century). In an expanding and competitive satellite industry, many companies are launching "better-faster-cheaper" satellites—and a key cost-saving measure is to scrimp on heavy and expensive shielding that protects satellites from the natural environment.

One would think that events like the May 1998 storm would convince satellite operators and their customers to spend more on their satellites or to prepare for more frequent losses. The lesson that the paging and telecommunications companies learned is that more backups are needed. The broadcast television industry, for example, is known to spread out its wireless connections and have alternates lined up if a satellite is knocked out. But many newcomers to the world of satellites—the cable and satellite TV companies, the upstart phone services, the teleconferencing networks, the financial markets—could be caught without an alternative when their satellite fails.

"It seems inadvisable to have such complex, societally significant systems susceptible to single spacecraft failures," notes Dan Baker. "With spacecraft operators working much closer to the margins, and with the recent record of space environmental disturbances, I believe many more catastrophic failures may occur."

In his testimony to a Senate panel in May 1998, computer scientist Peter Neumann of SRI International warned lawmakers about the dangers of too many eggs in one satellite basket. "The failure of Galaxy IV is just another example of the flaky infrastructure we're dealing with," Neumann told the Senate Committee on Governmental Affairs. "The companies that are smart have anticipated things like this and they have backups. But there are also a lot of naive folks who are dependent on one technology." He added: "As-yet-undiscovered vulnerabilities may be even greater than those that are known today. Future disasters may involve vulnerabilities that have not yet been conceived, as well as those that are already lurking."

2   Sun-Eating Dragons, Hairy Stars, and Bridges to Heaven

When beggars die, there are no comets seen;
The heavens themselves blaze forth the death of princes.
                    Shakespeare, *Julius Caesar*, Act 2.2

In the winter of 1504, Christopher Columbus had a problem. He was stranded with a mutinous, hungry crew and a creaky boat in St. Anne's Bay, Jamaica, in the midst of his fourth voyage to the New World. With their boat desperately in need of repair, the crew had beached the vessel, awaiting the return of assistance from Spain. They waited for nearly a year, but the wait was not the problem; the unruly crew was.

For much of the year, the local tribes on Jamaica had welcomed and bartered with Columbus and his men, at least until the rowdy sailors wore out their welcome. Alienated by the crew, the local tribes cut off the food supplies. The threat of mutiny, not to mention starvation, was rising in the European camp. Columbus needed a miracle, or at least a trick, to get out of trouble.

Checking his navigational tables, the explorer found that a lunar eclipse was predicted to occur on February 29, 1504. So he arranged a meeting with the native leaders for that evening. Columbus told the tribes that his God did not care for the way the sailors were being treated. To show his displeasure, God was going to take the Moon away. The native leaders watched with interest at sunset, as the Moon rose with the reddish glow of early, partial stages of the eclipse. As the lunar disk cleared the horizon, it appeared that a chunk had been taken out of its bottom. Within the hour, the entire Moon was blotted out by Earth's shadow.

As the story goes, the natives became terrified and offered to resume the supply of food to Columbus and his men. The explorer said he would think about it and took some time to confer with God, just enough time to return before the total eclipse ended 40-odd minutes later. God would pardon them, Columbus asserted, and bring back the Moon. And so he did. Food supplies were plentiful thereafter, and the ship's crew was a bit more contrite, at least until the rescue ship from Europe arrived.

Columbus had no direct line to a deity, of course, but he did have a keen insight into the human mind and a decent set of reference materials. He likely had information on the Saros, the ancient mathematical formula devised by the Babylonians to predict the cycles of solar and lunar eclipses. It was probably the most valuable cargo on his ship.

For thousands of years, humans have been intrigued by the cycles in the skies, from the daily rising and setting of the Sun and stars to the monthly phases of the Moon to the yearly procession of the seasons. In the human mind, the repetition of these cycles provides a hint and hope of order in the universe. Central to that order is the idea that events in the heavens—particularly lunar and solar eclipses—can alter or portend events on Earth.

In the name of both science and soothsaying, sky watchers from Persia, Greece, China, the Middle East, the ancient Americas, and Europe spent thousands of years developing the charts to predict future lunar and solar eclipses. Whether natural or supernatural, eclipses had an effect on civilization, and those who could predict eclipses had power in their hands. Even Stonehenge and the Medicine Wheels of the American prairies—well known as mystical calendars of the seasons—may have included mechanisms for predicting when Earth and Moon would cast their shadows on each other.

The word *eclipse* comes from a Greek word meaning "abandonment" or "to leave," and for many cultures a solar eclipse was seen as the Sun abandoning the Earth. Solar eclipses happen when the new Moon passes directly between the Sun and Earth, leaving part of the Earth in the Moon's shadow.[1] Total eclipses do not take place every time there is a new Moon because of the tilt and variation of the Moon's orbit. On average, the Moon passes directly in front of the Sun about once a year. Conversely, during a total lunar eclipse, the full Moon passes through Earth's shadow, and the Earth blocks all direct sunlight from the Moon.

The earliest record of a solar eclipse comes from ancient China around the year 2134 B.C.E. The ancient document *Shu Ching* records that "the Sun and Moon did not meet harmoniously." Chinese folk legends of the time held that an eclipse was actually the work of an invisible dragon devouring the Sun, prompting people to make loud noise during the eclipse in order to frighten the dragon and restore daylight. According to the story, the two royal astronomers, Hsi and Ho, failed to predict the event and properly warn the people. The astronomers were promptly executed.

There is little to no mention of notable eclipses in documents from other ancient cultures, though it seems that the ancient Babylonians were the first civilization to start working out the Saros cycles, the mathematical formula of the orbital position of the Moon and Earth that calculates when eclipses will appear over some geographic area. But there was a famous eclipse recorded during the biblical era. In the eighth century B.C.E., the Hebrew prophet Amos described an eclipse in his oracles denouncing the Jewish people. He wrote: "On that day, says the Lord God, I will make the Sun set at midday and cover the Earth with darkness in broad daylight." Whether Amos was referring to a specific day or a more apocalyptic vision, historical and scientific records show that there was a total solar eclipse on June 15, 763 B.C.E. An Assyrian historical record known as the Eponym Canon confirms the date, and a scribe at Nineveh also recorded this eclipse.

Perhaps the most famous solar eclipse of ancient times occurred in 585 B.C.E. The Greek philosopher Thales somehow predicted a solar eclipse for May 28, 585 B.C.E., though that prediction was not widely known at the time. The eclipse occurred in the midst of a battle between the nations of Lydia and Media. Seeing the day turned into night, the startled warriors stopped fighting and agreed to a peace treaty. They cemented the bond with a double marriage among the royalty.

The mythic power of the eclipse has at times been accentuated by connections to other natural disasters and earthly events. In the apocalyptic Book of Revelation, the Gospel writer John writes: "Then I watched as he broke open the sixth seal, and there was a great earthquake; the Sun turned as black as dark sackcloth and the whole Moon became like blood." Like Amos, John was writing symbolically, but he had tapped into an existing cultural belief—that eclipses and earthquakes are somehow linked. Years before, while writing about the Peloponnesian War, the Greek historian Thucydides noted that "earthquakes and eclipses of the Sun came to pass more frequently than had been remembered in former times." In another passage he wrote: "There was an eclipse of the Sun at the time of a new Moon, and in the early part of the same month an earthquake." Another Greek writer, Phlegon, reported that "in the fourth year of the two hundred second Olympiad, there was an eclipse of the Sun which was greater than any known before, and in the sixth hour of the day it became night; so that stars appeared in the heaven; and a great earthquake that broke out in Bithynia destroyed the greatest part of Nicaea."

More than a thousand years later, John Milton wrote in the epic *Paradise Lost*:

As when the Sun, new risen,
Looks through the horizontal misty air,
Shorn of his beams, or from behind the Moon,
In dim eclipse, disastrous twilight sheds
On half the nations and with fear of change
Perplexes monarchs

Some scholars assert that Milton was alluding to Emperor Louis I, son of Charlemagne. Shortly after witnessing the eclipse of May 5, 840, Louis died. Legend holds that he was so "perplexed" by the eclipse that he died of fright. The death became historically significant because it started a struggle for succession that ended with the Treaty of Verdun, dividing the Holy Roman Empire into the three major areas we know today as France, Germany, and Italy.

Tales of great eclipses and of their ominous, wondrous, or merely coincidental effects on civilization continued right up to the era of modern science. Millions of pilgrims and tourists still trek to the "path of totality"—the swath of Earth where the Sun is totally eclipsed in any given eclipse—as they did in 1998 and 1999 to the Caribbean, Europe, and the Middle East. Grand celebrations and cultural events are planned around the days when our Sun is taken away.

Even scientists celebrate eclipses, though not just for their mythic value. To a solar physicist, an eclipse is the only natural way to see the atmosphere (chromosphere and corona) of the Sun (see Figure 4). When compared with the density of the gas and the intensity of the light on the Sun's visible surface, or photosphere, the hot ionized gas (or plasma) of the upper atmosphere of the Sun is a million times dimmer. With the face of the Sun blocked by the Moon during a solar eclipse, the corona shines with the brightness of a full Moon. In fact, until French astronomer Bernard-Ferdinand Lyot developed a device known as a coronagraph (which uses a

FIGURE 4. Eclipses reveal the Sun's corona, the tenuous outer atmosphere composed of streams of energetic charged particles. With the brilliant disk of the Sun blocked, the faint light of the corona reveals streamers of solar wind blowing out into space. This eclipse was photographed in 1991 from atop Mauna Kea, Hawaii. Courtesy of the High Altitude Observatory.

disk to block the visible surface of the Sun) in 1930, eclipses marked the only time anyone could see, no less study, the corona.

The view provided by an eclipse or coronagraph is important because the corona is where the action is. The Sun's atmosphere is mysteriously millions of degrees hotter than the surface, and it is from this region that the solar wind originates, solar prominences appear and disappear, and giant bubbles of plasma—coronal mass ejections—grow and burst into the solar system. It is the corona, the tenuous atmosphere of our star, in which storms from the Sun arise.

Sun-Eating Dragons, Hairy Stars, and Bridges to Heaven

Like eclipses, comets have inspired panic and awe since ancient times. "The celestial phenomena called comets excite wars, heated and turbulent dispositions in the atmosphere, and in the constitutions of men, with all their evil consequences," wrote Claudius Ptolemy in the first century A.D. A Chinese silk book from the fourth century B.C.E. describes 27 types of comets and the specific calamities each produced. Aristotle wrote about comets in his *Meteorologica* (around 350 B.C.E.), and to the Greeks comets were known as "hairy stars." A few hundred miles to the west, the Romans saw one of these hairy stars as a sign of the divinity of Caesar. According to Plutarch, a comet appeared around the time of Julius Caesar's death in 44 B.C.E. Reports from China confirm a comet around that time and, 16 centuries later William Shakespeare made symbolic and poetic use of the comet in his tragic play about the Roman leader.

Many other comets have been associated with historic events. Halley's Comet glided across the skies in A.D. 66, just a few years before the fall of Jerusalem in 70. In A.D. 79 the volcano Vesuvius erupted and destroyed the cities of Pompeii and Herculaneum while a comet trailed across the sky. The passing of the centuries did little to squelch the fear of ominous comets. A French physician, Ambroise Paré, wrote that a comet in 1528 "was so horrible, so frightful, and it produced such great terror that some died of fear and others fell sick. It appeared to be of extreme length, and was the color of blood." In 1665 many were falling sick—the Plague was killing 90,000 people—as another great comet passed. Even during the American Civil War, the Great Comet of 1861 and comet Swift-Tuttle of 1862 were referred to as the "First and Second Civil War Comets."

The most famous and infamous of all comets, of course, is Halley's. Named for the English scientist who calculated its orbit and predicted its return, Halley's comet cruises across the sky every 76 years and has historically brought trouble with it. Known to

Chinese observers since at least 240 B.C.E., it may be the first comet ever recorded (some accounts suggest it may have been recorded as far back as 1059 B.C.E). Its first famous appearance came in A.D. 1066, on the eve of the Battle of Hastings, when William the Conqueror overcame his Anglo-Saxon foes in England. That appearance of Halley's comet was immortalized in the Bayeux tapestry. Four centuries later Pope Callixtus III excommunicated Halley's comet as an "instrument of the devil."

The return of Halley's comet from September 1835 to February 1836 was perhaps the most infamous, as the comet was blamed for many things. More than 500 buildings burned in New York City during a fire that blazed for several days. Along the border of Texas and Mexico, the comet presaged the sacking of the Alamo on March 6, or was it signaling the end of the reign of General Santa Anna? In South Africa, 10,000 Zulu warriors massacred 97 Boer men and women and 185 children at Weenen. In that year, wars erupted in Cuba, Mexico, Ecuador, Central America, Peru, Argentina, and Bolivia. Osceola, chief of the Florida Seminoles, allegedly prayed to the "Big Knife in the Sky" shortly before his warriors sacked Fort King.

Not everything about the 1835 return of Halley was bad. Legendary American author Mark Twain was born the year the comet crossed the skies. Later in his life Twain was known to say that he came in with the comet and would go out when it came again. The half-baked prophecy came true when Twain died in 1910 just before Halley returned.

It was too bad that Twain passed away before he could write about the comet's 1910 appearance, for the chicanery and foolishness surrounding the event were just the sort of tale he loved to tell. Astronomers had predicted that Earth would pass through the tail of the comet during May 1910, and some people panicked. Years earlier scientists had detected cyanogen, a poisonous gas, in the tail of a comet. That was just the opening the charlatans needed. As startled citizens spread rumors and ignorant proclamations, entrepreneurs made small fortunes selling "comet pills" that would counter the effects of the cyanogen gas that Halley would bring. But when the Earth passed through Halley's tail on May 20, no one died or even gasped. A few wallets were lighter.

Even as late as 1996 and 1997 comets were still ominous signs for some people. With the appearance of comets Hyakutake and Hale-Bopp, and with the apocalyptic atmosphere of the end of the millennium, comets again achieved notoriety. Supermarket tabloids suggested that Hyakutake was going to hit the Earth, and fringe cults and groups proclaimed that a UFO was trailing Hale-Bopp to take us out or to take some of us away. The hysteria culminated in the suicide of 39 members of the Heaven's Gate cult on March 26, 1997. The group believed that an alien spacecraft was hiding in the tail of comet Hale-Bopp, ready to take them to paradise.

Whether the passage of comets actually has an impact on civilization—or whether humans have created their own self-fulfilling prophecies—only the theologians and psychologists can answer. But these dusty ice balls do reveal the signature of at least one natural phenomenon that affects life in the solar system: the solar wind. The solar wind is a stream of electrically charged particles—essentially, hydrogen gas that is heated to a point where it is broken into its constituent protons and electrons, or plasma.[2] Every minute the Sun sprays millions of tons of this plasma in all directions at 1 million miles per hour. Yet the solar wind would not ruffle the hair on your head. In the vastness of three-dimensional space, the particles become so spread out that the solar wind has less mass per cubic centimeter (density) than even the best vacuums created in laboratories on Earth. As scientists learned in the twentieth century, it is the energy and the magnetic fields carried in that tenuous solar wind that cause space weather and blow pieces of comets away as they approach the Sun.

The first indication that the Sun might be emitting a "wind" came from comet tails. Kepler in the early 1600s guessed that the pressure of sunlight created those tails, and he was mostly right. The dust tail of a comet is usually bright and white, curling slightly as the Sun vaporizes part of the head of the comet. But each comet also has a second faint tail that stretches away from the Sun. The lightly colored (usually blue) "ion" tail can accelerate suddenly and can become distorted and kinked as super-fast streams of solar wind slough ions off of the comet. Both tails always point away from the Sun—whether the comet is inbound or outbound—as the ice and dust ball is buffeted by the outward-flowing solar wind and vaporized by the warming sunlight.

It is this solar wind—seen only in comet tails—that carries storms from the Sun to Earth. Our ancestors were not entirely wrong when they perceived some cosmic mischief at work. Comets don't bring peril to Earth, and they don't have much effect on life on the surface of Earth (except for those occasional collisions). But comets are shaped and affected by the Sun and the solar wind, making them more like kin than aliens.

To the dispassionate, objective viewer, auroras can appear as colorful, wispy curtains of light ruffling in the night sky. Sometimes the northern and southern lights (*aurora borealis* and *aurora australis*) stretch across the night as diffuse, flickering bands of green, blue, white, and red. Other times they streak the sky with rays or shafts of light. But for most of history, humans have seen a lot more in the heavens that just a brilliant, ghostly light show.

Aristotle, writing in his *Meteorologica*, made one of the first truly scientific accounts of the *aurora borealis*. The ancient Greek philosopher and scholar described "glowing clouds" and a light that resembled the flames of burning gas. He noted that these flames spread and at the same time sent out sparks and rays.

In Rome auroral arcs were first regarded as the mouth of a celestial cave. The Roman philosopher Seneca proposed that the auroras were flames slipping through cracks in the heavenly firmament. In A.D. 37, Emperor Tiberius saw the reddish light in the sky and thought the port of Ostia was burning. He dispatched troops to aid and rescue the inhabitants of the city, as noted by Seneca: "The cohorts hurried to the succor of the colony Ostia, believing it to be on fire. During the greater part of the night, the heaven appears to be illuminated by a faint light resembling a thick smoke." They marched to find nothing amiss.

Eighteen hundred years later, in 1839, all the fire brigades of London were sent north of the city to put out a fire that turned out to be an auroral blaze in the sky. A century later they were fooled again. "The ruddy glow led many to think half the city was ablaze," the Associated Press reported from London on January 25, 1938. "The Windsor fire department was called out in the belief that Windsor Castle was afire."

Years before Mediterranean peoples were chronicling the northern lights and Londoners were chasing them, the prophet Ezekiel may have experienced them as part of a spiritual vision. Around 593 B.C.E., during the Israelites' exile in Babylon, he wrote: "While I was among the exiles by the river Chebar, the heavens opened and I saw divine visions. As I looked, a stormwind came from the North, a huge cloud with flashing fire from the midst of which something gleamed like electrum. Within it were figures resembling four living creatures. . . . In among the living creatures something like burning coals of fire could be seen; they seemed like torches, moving to and fro. . . . The fire gleamed, and from it came forth flashes of lightning."

Other supernatural creatures have been spied in the dancing lights. Ancient folklore from China and Europe describe auroras, with their twisting, snake-like shapes, as great dragons or serpents in the skies. One theory holds that the dragon faced down by Britain's patron Saint George was in fact the aurora swirling in the sky over Scotland.

In the Nordic regions of Scandinavia, Iceland, and Greenland, the aurora was often seen as the great bridge Bifrost. In Norse mythology, this burning, trembling arch was the passage for the gods—particularly Thor, the god of war—to travel from Heaven to Earth. The Finnish parallel was the river of fire, Rutja, which marked the boundary between the land of the living and the dead. Tales from the Inuit around Hudson Bay echo the Nordic bridge legends: "The sky is a huge dome of hard material arched over the flat earth. On the outside there is light. In the dome, there are a large number of small holes, and through these holes you can see the light from the outside when it is dark. And through these holes the spirits of the dead can pass into the heavenly regions. The way to heaven leads over a narrow bridge which spans an enormous abyss. The spirits that were already in heaven light torches to guide the feet of the new arrivals. These torches are called the northern lights." Some other Native American Indian tribes envisioned spirits carrying lanterns as they sought the souls of dead hunters, while Eskimos saw souls at play, using a walrus head as a ball. The Fox Indians of Wisconsin feared the shimmering ghosts of their dead enemies. Some Vikings envisioned the Valkyries carrying torches—or reflected light off their shields—as they led slain warriors to Valhalla.

Those who did not see supernatural beings or *vigrod* ("war-reddening") often interpreted the aurora as a predictor of the weather. Snow and bitter cold were often thought to follow bright auroral displays in Scandinavia, while the Eskimos saw just the opposite: the spirits were bringing favorable weather. In an attempt at a physical explanation of the aurora—and the first description to invoke the name "northern lights"—an anonymous Norwegian author took a more sober and scientific approach. The author of the *Kongespeilet*—"The King's Mirror"—wrote in 1230 that: "These northern lights have this peculiar nature, that the darker the night is, the brighter they seem, and they always appear at night but never by day, most frequently in the densest darkness and rarely by moonlight. In appearance, they resemble a vast flame of fire viewed from a great distance. It also looks as if sharp points were shot from this flame up into the sky. . . . Otherwise it is the same with the northern lights as with anything else we know nothing about: that wise men put forward ideas and simple guesswork, and believe that what is most common and probable. Some people say that when the sun is under the horizon at night, some rays of light reach up to the skies over Greenland, a landmass so close to the edge of the earth that the earth's curvature in which hides the sun must be less there."

Some of the brightest minds in human history have puzzled over the aurora. Anders Celsius noted in his diary that volcanoes erupting near the North Pole might cause northern lights, as sulfur spewed into the atmosphere from the bowels of the Earth; later he claimed they were caused by reflected moonlight.[3] Benjamin Franklin attributed the lights to a sort of lightning and electric discharge from clouds and above the atmosphere, since he thought that electricity could not enter the ice of the polar regions.

Marten Triewald even made an attempt to create artificial auroras, using a shaft of light through a darkened room, a prism, and aquavit liquor. He was a persistent promoter of the idea—also espoused by René Descartes—that auroras were caused by the refraction of moonlight by the atmosphere and the reflection of different colored rays by ice crystals. In a paper entitled *"Experimentum aurorae borealis artificialis,"* published in the *Proceedings of the Royal Swedish Academy of Sciences,* Triewald described his attempt to simulate the play of light and vapor: "One observes with wonderment a northern light so natural that nothing can be more similar, and as the surface of the aquavit is quickly warmed by the colored sunbeam and in consequence evaporates, so one perceives most wondrous movements on the screen, on which flashing beams shoot suddenly up and then transform into colored veils, endlessly changing position between themselves . . . one sees all the phenomena that the natural northern lights display and as changeable as the same . . . it is never twice the same, just like the northern light."

We now know that the aurora is, in fact, a benign and beautiful sign that something electric is happening in the space around Earth. Named for the Roman goddess of dawn, auroras occur when fast-moving particles trapped in Earth's magnetic field come crashing down from space into the gases of Earth's upper atmosphere. Those electrically charged particles (electrons and protons) are governed by magnetic fields and can only move along Earth's invisible field lines. As the solar wind pours energy into the space around Earth and energizes the magnetic field, some of the trapped particles slide down the field lines and into the atmosphere, forming ovals of light centered on the north and south magnetic poles.

Because of the near-constant breeze of solar wind particles, auroral displays occur nearly every night in these ovals between latitude 60 and 70 degrees. (A common misconception is that auroras occur over the geographic poles. But, in fact, you would have to look south for an aurora if you were standing at the North Pole.) Nightly light shows are one of the privileges of living in the frigid extremes of Canada, Alaska, Scandinavia, Scotland, and Russia and at the far edges of New Zealand, Chile, and Antarctica.

The auroras in the northern and southern skies are quite nearly mirror reflections of each other, or "conjugates" as scientists call them. The first recorded sighting of conjugate auroras occurred in September 1770, during the expeditions of British Captain James Cook. While exploring Australia and the South Pacific on the *HMS Endeavour*, Joseph Banks, the crew's naturalist, noted: "A phenomenon appeared in the heavens in many things resembling the *Aurora borealis*." Later studies of the *Qing-shigao*, a draft history of the Qing Dynasty of China, revealed that an aurora was observed on the same night—September 16, 1770—in the northern hemisphere. As recently as October 2001, scientists gathered images of auroras occurring simultaneously in the northern and southern hemispheres, confirming that the auroral ovals mimic each other.

During more intense periods of space weather (when a coronal mass ejection or an intense flare alights from the Sun), the auroral ovals descend to lower latitudes, bringing the northern lights to such cities as Boston, Seattle, Minneapolis, or Edinburgh (about 15 to 20 times per year). Perhaps once per solar cycle the aurora can be viewed most of the way toward the equator, as it was in 1909 when the most potent magnetic storm on record brought an aurora to Singapore. These "Great Auroras" as scientists call them are the sort that inspired Norse warriors and Mediterranean philosophers, the kind that led campers in the Appalachian Mountains to believe a nuclear war had begun in March 1989. They are provoked by the severest of storms from the Sun, and for most humans they are the only visible manifestation of a space weather event.

To the naked eye, the face of the Sun appears to be unblemished, constant, and pure. And from the time of Aristotle that was the culturally and politically correct point of view, at least in Western civilization. The heavens were perfect and unchanging, just as the gods, and then God, had made them. A spot on the Sun would mean that there is change and impurity in the heavens, and nothing could be more constant and pure than the Sun.

Yet those who watched closely, those who squinted through the thick haze or at the sunset, those who were willing to risk blindness for a peek at our star definitely saw sunspots. The oldest known records of spots on the Sun come from China in 28 B.C.E., and there is some evidence that Chinese astronomers may have seen them years before. The Greek philosopher Anaxagoras may have observed a spot in 467 B.C.E., and Theophrastus may have spied another one in the fourth century B.C.E.

But given the dominance of Aristotle's cosmology—which was later adopted by the Catholic Church—there are few other records of any other sunspots viewed in Western civilization until the time of Galileo. Sunspots were deemed to be physically impossible, and most appearances of black splotches on the face of the Sun were explained away as planets—primarily Mercury and Venus, and the mythical planet Vulcan—passing in front of the Sun.

One of those few accounts was composed in A.D. 1128, when an English monk sketched the oldest known drawing of a sunspot. After spying a pair of sunspots on December 8, John of Worcester drew a picture of the Sun's blemished disk and wrote: ". . . from morning to evening, appeared something like two black circles within the disk of the Sun, the one in the upper part being bigger, the other in the lower part smaller." The fact that he could see the sunspots with the naked eye and that he could make out the umbrae and penumbrae of the spots suggest that they must have been extremely large.

Sunspots became a lot easier to see, and much harder to ignore, with the invention of the telescope in 1609. By the following year, astronomers had trained their new eyes on the Sun and began the first detailed studies. Within the course of two years, four men independently and almost simultaneously confirmed the existence of sunspots: Johannes Fabricius in Holland, Thomas Harriot in England, Galileo Galilei in Italy, and the Jesuit mathematician Christopher Scheiner in Germany. Officially, Harriot made the first observation, as recorded in his notebooks on December 8, 1610, but he did not follow those observations with any serious studies. Fabricius detected sunspots in March 1611 and quickly became the first to publish his observations. His book, *De Maculis in Sole Observatis* (*On the Spots Observed in the Sun*), was published in fall of 1611; yet none of the other three astronomers were aware of his publication for many years.

While Harriot and Fabricius recorded the official firsts, Galileo and Scheiner made the first truly scientific studies of sunspots and began to infer the physical properties of this new Sun with spots. Galileo claimed to have first seen sunspots late in 1610, but he did not actually write much about them until 1613, when he wrote the first of several responses to Scheiner's theory that the spots were actually moons or planets with orbits very close to the Sun.

Shunning Scheiner's interpretation—and the Church-sanctioned Aristotelian view of a perfect heaven—Galileo argued that spots were a feature of the surface or atmosphere of the Sun. He could not say what they were, but to Galileo the spots looked like clouds (see Figure 5). Ultimately, it was this assertion that the Sun had spots—combined with the discovery of Jupiter's moons and the promotion of Copernicus's Sun-centered view of the solar system—that led to Galileo's excommunication from the Roman Catholic Church.

Given such a philosophical and cultural climate, extensive studies of sunspots were mostly kept in the closet until the eighteenth century. By then solar science was thriving and astronomers began keeping daily logs of the number of spots on the Sun. The first regular observations began in 1749 at the Zurich Observatory in

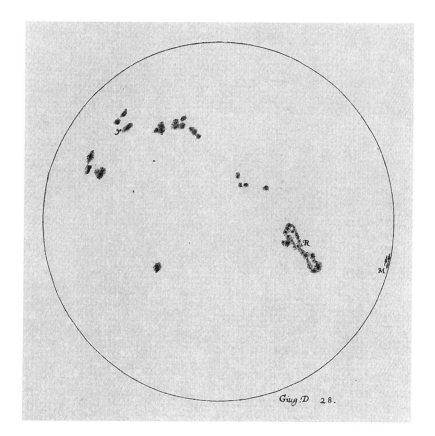

FIGURE 5. Galileo sketched this image of the face of the Sun on June 28, 1613. By collecting observations at about the same time each day, he and other sunspot watchers were able to decipher the motion of the spots across the solar disk. Courtesy of Owen Gingerich/Harvard-Smithsonian Center for Astrophysics.

Switzerland, and with the addition of other facilities in other parts of the world, continuous record keeping started in 1849.

An amateur astronomer, Heinrich Schwabe, was the first to turn sunspot sketches and records into a scientific advance. He studied the Sun day by day in search of the mysterious planet Vulcan, which many sky watchers believed to be the closest planet to the Sun, inside the orbit of Mercury. After 17 years of close observation of the Sun, he noted in his "Excerpts from Solar Observations During 1843" (*Astronomische Nachrichten*) that there was actually a cycle, a rhythm, to the appearance of sunspots: "The weather throughout this year was so extremely favorable that I have been able to observe the Sun clearly on 312 days; however, I counted only 34 groups of sunspots. . . . From my earlier observations, it appears that there is a certain periodicity in the appearance of sunspots, and this theory seems more and more probable from the results of this year." The number of sunspots rose and fell in a definite pattern, Schwabe noticed, over the course of his 17 years. Recalling his ill-fated search for a planet that did not exist, Schwabe said: "I may compare myself to Saul, who went to seek his father's ass and found a kingdom."

The initial publication of Schwabe's work did not draw much attention. But eventually scientist and explorer Alexander von Humboldt discovered the paper. He published Schwabe's table in his encyclopedic compilation of natural science, *Cosmos*. Suddenly, scientists around the world became interested in the sunspot cycle, reconstructing older records and comparing the changes in sunspot numbers with the number of magnetic storms detected on Earth. When scientists examined longer spans of solar observations, they found that the number of sunspots rose and fell in a cycle that lasted about 10 to 12 years. And by comparing 20 years of magnetic field data with Schwabe's sunspot data, English scientist and general Edward Sabine announced in 1852 that he had found a pattern in the occurrence of magnetic storms that paralleled the rise and fall of sunspots.

Schwabe had been looking at the Sun to discover a planet inside the orbit of Mercury. Instead he discovered the most fundamentally important trait of our Sun: it is dynamic and changeable, and those changes are cyclical. It set the stage for successors to discover how changes in the appearance of the Sun produced changes on the Earth.

○●○

Bound to the Earth, our only naturally occurring experience with space weather comes from what we can see with our eyes: eclipses, comets, auroras, and sunspots. And since our vision is distorted by Earth's atmosphere and limited to rays of visible light, it is easy to understand how we have turned eclipses and comets into divine portents and auroras and sunspots into inexplicable curiosities. When you consider that phenomena such as comets and auroras are influenced by invisible force fields (gravity and magnetic field lines) and tenuous gases of pure atomic particles (plasmas), it seems easier to believe in mythical auroral spirits, UFOs hiding behind comets, and earthquake-inducing eclipses than in space weather.

In trying to understand and explain the patterns and quirks of nature in spiritual and poetic terms, our ancestors were on to something. Eclipses, comets, auroras, and sunspots are indeed wondrous portents and signs, and the heavens really do affect life on Earth. As signals of space weather, these phenomena affect how we live in a modern technological civilization, as scientists learned firsthand in 1859.

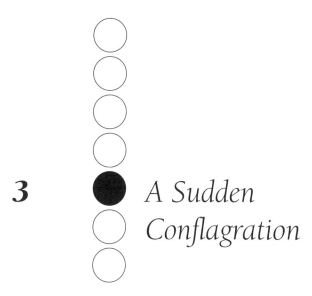

3 A Sudden Conflagration

It is not impossible to suppose that in this case
our luminary was taken in the act . . .
                              Balfour Stewart

On September 1, 1859, Richard Carrington was doing what he did every day: he observed the Sun. The son of a prosperous English brewer, Carrington was more interested in astronomy than in the theology he originally set out to study. He had built his own observatory in Redhill, England, and had dedicated part of each day to observing the stars and the Sun. His research had been fruitful, as he discovered that sunspots tend to occur at different latitudes on the Sun during different stages of the Sun's activity cycle (the cycle itself was just being uncovered by Schwabe, Sabine, and von Humboldt). By chronicling the day-to-day progress of sunspots, Carrington also found that the Sun rotates faster at the equator than it does at higher latitudes.

But at 11:18 a.m. on that late summer day in 1859, Carrington made perhaps his most compelling discovery. He had been projecting the image of the sun onto a screen—the only safe way to observe the Sun with a telescope—and sketching the sunspots as they appeared in the image. Suddenly, he noticed bright patches of white light around some of the spots in the Sun's northern hemisphere. Thinking that his telescope or filter was allowing light to leak into the projected image, he adjusted his instrument. But the "sudden conflagration" of white blotches did not go away.

Carrington rushed from the room to get a witness, but by the time he got back, the white flash was gone. Another hour of observation left him disappointed, as the bright patch never reappeared. Months later Carrington described the event in the *Monthly Notices of the Royal Astronomical Society*:

> While engaged in the forenoon of Thursday, Sept. 1, in taking my customary observation of the forms and positions of the solar spots, an appearance was witnessed which I believe to be exceedingly rare. . . . I had secured diagrams of all the groups and detached spots, and was engaged at the time in counting . . . when within the area of the great north group (the size of which had previously excited general remarks), two patches of intensely bright and white light broke out. . . . I thereupon noted down the time by the chronometer, and seeing the outburst to be very rapidly on the increase, and being somewhat flurried by the surprise, I hastily ran to call some one to witness the exhibition with me, and on returning within 60 seconds, was mortified to find that it was already much changed and enfeebled. Very shortly afterwards the last trace was gone. . . .

Lacking a witness from his own observatory, Carrington was fortunate to find another objective observer down the road. Fellow sun watcher Richard Hodgson had observed the same conflagration from his observatory in Highgate, England. He wrote in the *Monthly Notices*:

> While observing a group of solar spots on the 1st September, I was suddenly surprised at the appearance of a very brilliant star of light, much brighter than the sun's surface, most dazzling to the protected eye, illuminating the upper edges of the adjacent spots and streaks, not unlike in effect the edging of the clouds at sunset; the rays extended in all directions; and the centre might be compared to the dazzling brilliancy of [a] bright star. It lasted for some five minutes, and disappeared instantaneously about 11.25 a.m.

What Carrington and Hodgson had witnessed was the first known sighting of a solar flare, a sudden and intense heating of the solar atmosphere comparable to an explosion. In fact, what they saw was an extremely rare "white light" flare. Most flares can only be observed in certain wavelengths with optical filters that had not yet been developed in Carrington's day. The fact that the flare was visible in white light, against the brilliant background of the white-hot Sun, suggests that it would have been monumental even if it had not been the first one observed. By Carrington's own accounting, the "two patches of light traversed a space of about 35,000 miles" across the face of the Sun (see Figure 6).

In the moments after the solar flare, Carrington was surprised to see that little else on the Sun had changed. "It was impossible not to expect" some change in the appearance of the spots on the Sun, he noted, but there were none. That meant that the flare must have been at least partly independent of the sunspots. "The impression left upon me is that the phenomenon took place at an elevation above the general surface of the Sun and, accordingly, altogether above and over the great group in which it was seen projected."

But the astronomical event did not end with the flare; in fact, the pyrotechnics were just starting. While Carrington had been

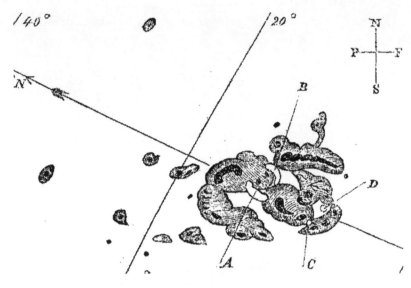

FIGURE 6. Richard Carrington sketched this picture of the great white light solar flare of September 1, 1859. Note the white bands buried amid the black and gray sunspots, depicting the twin bands of the flare as it burst into view around the sunspots. Sketch from the *Philosophical Transactions of the Royal Society of London*.

making his observations, the magnetic instruments at the Kew Observatory in London had been recording a distortion in the magnetic field of Earth. About 18 hours later, when the blast of plasma from the Sun reached Earth, the Kew magnetometer went wild. Earth was in the throes of one of the strongest magnetic storms ever recorded—a storm that lasted more than six days and zapped the most advanced communications system of the day, the telegraph.

The X rays from the flare bombarded the Earth's atmosphere almost instantly, cooking the ionosphere and producing a surge of electric currents. The solar flare was almost certainly accompanied by a coronal mass ejection (a phenomenon that was not actually discovered until the 1970s), a blast of hot electrified particles that sped from Sun to Earth at 2,300 kilometers per second (more than 5 million miles an hour). The shock wave and cloud smashed into the Earth's magnetic field, causing a huge increase in the flow of invisible electric currents in space and in our atmosphere. Those currents were strong enough to affect the strength of Earth's magnetic field, as detected on the ground; scientists call it a magnetic storm.

In the decade leading up to Carrington's flare, scientists such as General Edward Sabine had begun to suspect that solar activity could increase the auroral activity and could induce magnetic storms. So when Carrington learned in September 1859 that a magnetic storm had coincided with his flare, he came to suspect a physical connection between Sun and Earth. But in his notes to the Royal Astronomical Society, he qualified that connection by saying "one swallow does not make a summer." There were too few data—just his one flare—to make such a direct connection.

Others were less cautious about the connection between Sun and Earth. Another British scientist, Balfour Stewart, spent much of 1859 through 1861 collecting anecdotes and data from science-minded colleagues across Europe and the rest of the world. Stewart was inspired by the amazing aurora that was "observed very widely throughout our globe" and by the "magnetic disturbances of unusual violence and very wide extent." Stewart collected magnetic field measurements from several observatories but particularly from Sabine, who had concluded seven years earlier that the pattern of magnetic storms around Earth tracked closely with the sunspot cycle. Surely the September 1 and 2 events were a direct observation and test of that geophysical connection. In a paper published in the *Philosophical Transactions of the Royal Society* of London, Stewart resolved to be straightforward about the link: "The interest attached to these appearances is, if possible, enhanced by the fact that at the time of their occurrence a very large spot might have been observed on the disk of our luminary—a celestial phenomenon which we have grounds for supposing to be intimately connected with auroral exhibitions and magnetic storms. . . . If no connection had been known to subsist between these two classes of phenomenon, it would, perhaps, be wrong to consider this in any other light than a casual coincidence; but since General Sabine has proved that a relation exists between magnetic disturbances and sun spots, it is not impossible to suppose that in this case our luminary was taken in the act."

Across the Atlantic, Yale University professor Elias Loomis was simultaneously studying and analyzing that link between Carrington's sunspots, the flare, and the intense auroras on Earth. Loomis viewed the aurora from Lewiston, Maine. In the first of nearly a dozen papers he compiled on the events of September 2 for the *American Journal of Science and Arts,* Loomis noted that the auroral light show was "one of the most remarkable ever recorded in the United States . . . not only for the great extent of the territory over which it was observed, but also for the duration, for the intensity of illumination as well as the brilliancy of colors, and the extreme rapidity of the changes."

The storm generated a great deal of interest in the Americas. Ordinary people had observed the unusual displays with fascination, and many recorded what they saw. Joseph Henry, the first secretary of the nascent Smithsonian Institution, published a plea asking that the flood of letters reporting the event be directed to Loomis. Seeing the event as an opportunity to tease out the mysterious physics of the aurora and of Earth's global magnetic field, Loomis appealed to his colleagues to report the shape, duration, and variability of aurora: "It is of the highest importance to science that we should ascertain what the aurora is." By compiling and trying to make sense of the numerous accounts from around the world, Loomis intended to get a global picture of the event. He was gathering all the facts "in the expectation that at some future day they may afford the basis for a complete and satisfactory theory" of what causes the aurora.

And the reports flooded in from all over the world. Loomis and colleagues published dozens of accounts of the great aurora of September 1859. Observers wrote of auroras visible from the usual North American outposts—Toronto, New Haven, Halifax, and West Point—and from surprised viewers in Honolulu, St. Louis, San Francisco, New Orleans, Galveston, and Key West. Daniel Kirkwood of Bloomington, Indiana wrote: "The whole visible heavens were illuminated, the light at times being such that ordinary print could be read without much difficulty." In the southern hemisphere, observers reported auroral lights in Santiago and Concepción, Chile, and from all over Australia. And with help from Stewart and other colleagues in Europe, Loomis acquired data and descriptions from Rome, Athens, and Russia.

The most unusual reports of auroras came from the Caribbean. Andreas Poey, director of the Physio-Meteorological Observatory at Havana, Cuba, noted: "The appearance of the *aurora borealis* in the twenty-third degree of latitude is so rare that it naturally produces fear in the common mind, and arrests the attention of men of science." Though no one reported on the fears and superstitions of the common man on that night, plenty of descriptions from arrested scientific minds were filed from Puerto Rico, Jamaica, Bermuda, the Bahamas, and San Salvador. One U.S. naval officer even reported seeing the aurora from his ship off the west coast of Nicaragua.

Along with the many reports of fire in the sky, Loomis sought and received numerous reports of pyrotechnics in the telegraph offices around the world. Shortly after the first telegraph lines were established in the 1830s and 1840s, operators noticed that their systems behaved erratically when auroras were visible overhead. To send a telegraph message under normal conditions, operators manipulated and interrupted electric currents flowing through copper wires, starting and stopping with specific timings and sequences that made a code (such as Morse code). During auroral light shows, extraneous electric currents would flow through the wires, superseding the normal telegraph currents and making transmission of messages almost impossible. In one of the first scientific studies of these "spontaneous electrical currents," W. H. Barlow wrote: "On the evening of the 19th of March, 1847, a brilliant aurora was seen, and during the whole time of its remaining visible, strong alternating deflections occurred on all the instruments. Similar effects were observed also on the telegraphs on several other lines of railway." What Barlow and other operators did not know was that the auroral display was actually an indicator of strong electric currents flowing through the atmosphere near Earth's surface, playing havoc with Earth's magnetic field and pumping geomagnetically induced current into their lines.

Such was the case in September 1859. Loomis wrote in his first report that the widespread aurora of September 2 was "equally remarkable for the disturbances which accompanied it." Telegraph communication came to a standstill in many parts of North America and Europe. "These electrical perturbations were recorded not only by magnetic instruments, but also over the whole system of telegraph wires," wrote Loomis. "The magnetic induction either greatly interfered with or prevented the working of the lines . . . while in more than one case [the lines worked] solely by the atmospheric influence!"

O. S. Wood, superintendent of the Canadian telegraph lines, reported that he had never witnessed anything like it in 15 years working with telegraphs. "Well-skilled operators worked incessantly . . . to get over, in even a tolerably intelligible form, about four hundred words of the steamer *Indian*'s report for the press; but . . . so completely were the wires under the influence of the *aurora borealis* that it was found utterly impossible to communicate between the telegraph stations."

George B. Prescott, telegraph superintendent in Boston, told Loomis that the wires from Boston to Portland, Maine, were loaded with electric current all day, even though the system's batteries were disconnected. Clever operators decided to use the mysterious currents to their own advantage. Prescott described the circumstances:

> Upon commencing business at 8 o'clock a.m., it was found that all the wires running out of the office were so strongly affected by the auroral current as to prevent any business from being done. . . . At this juncture it was suggested that the batteries should be cut off, and the wires simply connected to the earth. The Boston operator accordingly asked the Portland operator to cut off his battery and try to work with the auroral current alone. The Portland

operator replied, "I have done so. Will you do the same?" The Boston operator answered, "I have cut off my battery and connected the line to the earth. We are working with the current from the *Aurora Borealis* alone. How do you receive my writing?" "Very well indeed," rejoined the Portland operator, "and much better than with the batteries on. There is much less variation in the current, and the magnets work steadier. Suppose we continue to work until the aurora subsides?" "Agreed," said the Boston operator.

The extra current also caused some unexpected electrical hazards for telegraph operators. In a report from Springfield, Massachusetts, observers noted that flames were seen shooting from the break-key of the telegraph to the iron frame, and the "heat was sufficient to cause the smell of scorched wood and paint to be plainly perceptible." In Washington, D.C., the smell was that of burned flesh, as telegraph operator Frederick Royce was zapped by his own equipment. "During the auroral display, I was calling Richmond and had one hand on the iron plate," Royce explained. "Happening to lean towards the sounder, my forehead grazed a ground wire. Immediately, I received a very severe electric shock. . . . An old man who was sitting facing me said that he saw a spark of fire jump from my forehead to the sounder."

In all, more than 70 reports were sent to the *American Journal of Science and Arts*, and perhaps dozens more were published or presented in Europe. The occasion of the first recorded solar flare had turned into the first widely observed instance of space weather, as the most advanced communication technology of the day had proved vulnerable to a blast from the Sun. Earlier observers had detected, individually, different aspects of the electric and magnetic connection between Sun and Earth. But never before had so many scientists and amateur observers witnessed and chronicled the flow of energy from the surface of the Sun to the surface of the Earth and into human technological systems.

"If it be true that the spots on the surface of our luminary are the primary cause of magnetic disturbances," Balfour Stewart wrote in the conclusion of his report on the September 1859 event, "it is to be hoped . . . that ere long something more definite may be known with regard to the exact relation between these two great phenomena." Indeed, it would not be long.

4 Connecting Sun to Earth

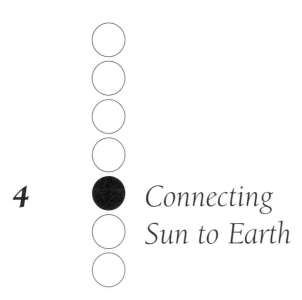

Some say that the Northern Lights are the glare of the Arctic ice and snow;
And some say that it's electricity, and nobody seems to know.
But I'll tell you now—and if I lie, may my lips be stricken dumb—
It's a mine, a mine of the precious stuff that men call radium.
                    Robert W. Service, "The Ballad of the Northern Lights"

In the years leading up to Carrington's flare and Loomis's auroral show, scientists had slowly become aware that one of the connections between the Sun and Earth might be magnetism. Successive experiments with the invisible force that attracts and repels objects and guides electricity forced scientists to consider that what they could see on the Sun and in the night skies was only half as important as what they could not see.

Around A.D. 1000, Chinese inventors used the mysterious magnetic properties of lodestone to develop the compass, and within a few hundred years, floating lodestones and magnetized needles pointed the way to the north and south for sailors and scientists. But it wasn't until a personal physician to Queen

Elizabeth I, William Gilbert, examined the subject that anyone made a systematic attempt to explain why a compass worked. In the late sixteenth century, Gilbert compiled the wisdom of his predecessors and began his own series of experiments and demonstrations with magnets, a few of which were eventually performed in front of the Queen. He took a spherical magnet that he called a *terrella*—little Earth—and moved a small compass around its surface. Wherever he moved it, the needle continued to point to the sphere's magnetic north pole. This led Gilbert to propose that the Earth itself was a giant magnet, with north and south poles, as if a great bar magnet had been buried inside. He published his ideas in 1600 in a treatise entitled *De magnete: Magnus magnes ipse est globus terrestris*, or *On the Magnet: The Earth Itself Is a Great Magnet*. It was the beginning of the modern era of geomagnetism and geophysics.

More than a century after Gilbert's exposé on magnets, a London watch and scientific instrument maker found that an unseen force was tinkering with his magnetized needles. In 1722, George Graham was measuring the deviation of the compass needle from true north.[1] He noticed that on some days the needle made irregular motions that he could not explain. The needle danced even when the compass was kept stationary inside of a glass box. Graham, who was a fellow of London's Royal Society, published his mysterious observations in the *Philosophical Transactions of the Royal Society*, but he had no satisfactory explanation.

Seventeen years later Graham (in England) and Anders Celsius (in Sweden) simultaneously detected the same sort of unexplained, irregular deflections of the compass needle. Celsius and his assistant, Olof Hiorter, noted that the deflection seemed to occur when auroras danced in the sky. Though the two men did not have a name for their phenomenon—Alexander von Humboldt would not devise one until the 1830s—Graham and Celsius had made the first direct observation of a magnetic storm.

The scientific understanding of electromagnetism is the great triumph, and the most important event, of the nineteenth century.

The pioneering physics experiments of Charles Augustin de Coulomb, Hans Christian Oersted, André-Marie Ampère, Michael Faraday, Carl Friedrich Gauss, Wilhelm Weber, and Joseph Henry uncovered the laws that link electricity and magnetism. They learned that changing magnetic fields produce electric currents and electric currents produce magnetic fields. The brilliant James Clerk Maxwell (who is buried close to Charles Darwin and Isaac Newton in Westminster Abbey) would develop his theory of electromagnetism, including the crown jewel—that light is an electromagnetic wave. The discovery and control of electricity and magnetism were as important to civilization as the discovery of fire.

Part of the exploration was led by such promoters of science as Alexander von Humboldt and General Edward Sabine, who set up global networks of magnetic observatories. Starting in the 1830s, Sabine, Gauss, Weber, and von Humboldt launched the "magnetic crusade" to tease out the secrets of Earth's magnetic field. They began charting the daily magnetic disturbances that Graham and Celsius discovered. The worldwide scope of these magnetic storms suggested that the environment was reacting to something grand in the space around Earth.

By 1859 there were enough magnetic observatories and interested observers to allow Loomis and Stewart to chronicle so precisely what we now know was the first great space weather event. In the years following, scientists made the first comprehensive global attempt to fit all the pieces of the space weather puzzle into one cohesive picture. And it was a catalyst for more extensive studies of the physics of the Earth—geophysics—and the space around it—space physics.

Much of the work over the next 40 years was centered on improving the observations, with both better instruments and systematic standardized measurements. For instance, American astronomer George Ellery Hale invented the spectroheliograph in 1892, a device that allowed him to view the Sun in individual wavelengths of light. By July 15 of that year, Hale was able to photograph the evolution of a large solar flare—a flare that preceded a magnetic storm about 19 hours later. He also used his new instrument to determine that sunspots were intensely magnetic.

By the end of the nineteenth century, the evidence was mounting that the Sun was affecting the magnetic field around Earth and stirring up storms and auroras. Many magnetic storms occurred in tandem with reports of solar flares or of large sunspot regions near the Sun's equator. Other times, as scientist E. W. Maunder discovered, storms recurred in 27-day periods, a span of time equal to the rotation rate of the Sun. Maunder boldly asserted in a paper to Britain's Royal Astronomical Society that "our magnetic disturbances have their origin in the Sun." But no one had a satisfactory explanation for the physics of the connection; namely, how and why the Sun should impact Earth in this way. And not all were convinced. The influential president of the Royal Society of London, Lord Kelvin, wrote in 1892 that "the supposed connection between magnetic storms and sunspots is unreal."

In Norway, physicist Kristian Birkeland used theory, experiment, and observations to try to provide a comprehensive physical explanation for the electromagnetic link between the Earth and the Sun. Birkeland had been a student of Henri Poincaré, one of the leading physicists of his time. He had done noteworthy theoretical work on electromagnetism, producing the first general solution of Maxwell's equations. In fact, one of Birkeland's papers—published in the journal *Comptes Rendus* in 1894—was referenced in the *American Journal of Physics* as recently as 1982. Nearly 100 years later, Birkeland's pioneering work was still relevant to physics.[2]

But Birkeland's true scientific love was the aurora, and he made it the central focus of much of his research. Reviving William Gilbert's experiments, Birkeland placed a magnetized sphere representing Earth—his own *terrella*—inside a vacuum chamber and fired cathode rays at it (see Figure 7). From centuries of reports and firsthand experience, Birkeland knew that auroras were mostly a phenomenon of the polar regions.[3] So Birkeland was not necessarily surprised when he fired cathode rays at the *terrella* and

FIGURE 7. Kristian Birkeland works in his laboratory to simulate the aurora by shooting beams of electrons at his *terrella,* or "little Earth." The apparatus could generate up to 20,000 volts. Photo from *The Norwegian Aurora Polaris Expedition, 1902–1903,* by Birkeland, published in Christiana, Norway, in 1908.

saw that the beam followed the magnet's field lines and hit the sphere near the poles. He surmised that the Sun must have been shooting beams of corpuscles (what we now call electrons) toward Earth, where the planet's magnetic field pulled them in near the poles. His young colleague, mathematician Carl Stoermer, added that since those "corpuscles" carried electric current, the magnetic field of Earth would be disturbed, as in a magnetic storm.

Birkeland's experiments failed to account for one of the most important traits of auroras: they are common around the polar regions but exceedingly rare at the poles themselves. Essentially, auroras form a ring around the poles but do not form over the poles. In a few experimental instances, Birkeland's beam produced a ring of light and what appeared to be a dark hole near the poles of his *terrella.* But mostly he observed just a cap of light on his "little Earth." Stoermer analyzed the motion of the electrons mathematically, but he could not find a compelling reason why electrons from the Sun should avoid the poles themselves.

Birkeland's theory of electron beams from the Sun had other problems. The repulsion between the like-charged electrons would quickly expand the beam; by the time the beam reached Earth it would be too diffuse to cause magnetic storms. Critics also pointed out that if the Sun got rid of many electrons via Birkeland's beam, it would soon become positively charged, and the negative electrons would not be able to escape. At a time when scientists knew about electrons—but the proton had not yet been discovered—Birkeland could give no satisfactory answer. Yet he refused to give ground on his theories. Birkeland's observations led him to other remarkable conclusions. In particular, he argued strenuously that electrical currents from space (carried by his electron beams) flowed down along Earth's magnetic field horizontally for a few hundred kilometers in the atmosphere and then back out to space.

Finally, Birkeland's theory of electron beams failed to account for the occurrence of magnetic storms close to Earth's equator. Again, Stoermer searched for an answer. In studying the behavior of electrons in a magnetic field, he found that some particles would reach the *terrella* near the poles, while others would become trapped around the equator, hovering at some distance from the little Earth. The electrons never actually reached the *terrella* at the equator, so how could Birkeland's electron beams from the Sun cause havoc at midlatitudes?[4]

Birkeland continued to pursue his ideas, but over time the quality of his work declined. Alex Dessler, a pioneering space physicist and Birkeland enthusiast, has suggested that Birkeland suffered from mercury poisoning, a common hazard for laboratory scientists of the time. Birkeland died in Japan in 1917, eight months short of his fiftieth birthday. His untimely death short-circuited the Nobel Prize for which he was being considered.

Just as Birkeland exited the scene, Sydney Chapman entered it. A brilliant theorist, Chapman made several seminal contributions to space physics, but he did not think much of Birkeland's ideas. He began to investigate magnetic storms following his own instincts and approaches. In 1930, Sydney Chapman and colleague Vincent Ferraro thought they had the best explanation for how the Sun could stir up the atmosphere and magnetic field of Earth. The two scientists postulated that magnetic storms and auroras were caused by clouds of electrically neutral gas, with equal amounts of negatively and positively charged particles; we now call this mixture of electrons and protons the fourth state of matter, plasma. Chapman and Ferraro argued that these clouds of particles were ejected from the Sun during a solar flare, flew across empty interplanetary space, and enveloped the Earth. These clouds would be excellent conductors of electricity and so would generate currents and distort Earth's magnetic field. Looking at more than a hundred years of magnetic observatory data, Chapman and Ferraro noted that most magnetic storms began with a sudden increase in the strength of Earth's magnetic field worldwide. That jump, they proposed, heralded the arrival of the cloud from the Sun.

But the plasma clouds from the Sun weren't just smacking directly into Earth. Chapman and Ferraro proposed that Earth's magnetic field carved a bubble or "cavity" out of space, one that would deflect and repel much of the spray from the Sun. And somehow this interaction of the Earth's magnetic field (dubbed the magnetosphere by geophysicist Thomas Gold in 1959) would cause some particles to be trapped around the Earth in a ring of electric current around the equator, though they could not explain why. Though some of the details were incorrect, the "Chapman-Ferraro cavity" formed the basis for the modern concept of space weather.

Although Chapman was right about many things, he was not necessarily right to dismiss many of Birkeland's theories. Swedish physicist Hannes Alfvén kept many of Birkeland's ideas alive, often in defiance of the conventional scientific wisdom. He saw merit in Birkeland's direct electrical connections between Sun and Earth and he updated many of those theories. Alfvén in his own right made several important theoretical discoveries in plasma physics (the physics of ionized gases). As with Birkeland, Alfvén's theories were not always accepted at first but once understood became crucial for understanding the physics of plasmas on Earth and in space.

And in the end he saw the scientific community verify Birkeland's idea of electric currents flowing from space down into the ionosphere and back out again. In 1970, Alfvén won the Nobel Prize for his work.

Pursuing and analyzing Chapman and Ferraro's theories, other observers pointed out that the Sun wasn't just ejecting intermittent plasma clouds. In fact, there was a steady breeze. Looking at the tails of comets, German scientists Cuno Hoffmeister and Ludwig Biermann proposed in the 1940s and 1950s that the Sun was constantly emitting something extra, something more than just light and occasional plasma blobs. They noted that comets had two tails—one of dust, one of ions—and only the dust tail could be explained by the pressure of sunlight pushing against the comet. The ion tail was not a steady, constant stream like the dust tail; it wiggled and bent regardless of the direction of the comet. Occasionally, the ions accelerated in a burst, though the speed of the comet had not changed. Hoffmeister and Biermann advanced a theory that the Sun was emitting a steady stream of particles, a "solar corpuscular radiation." These streams of plasma from the Sun, ebbing fast and slow, must be sloughing ions off of the comets. Of course, like the ring current around Earth, no one could explain why this radiation should exist.

In trying to decipher the structure of the corona of the Sun (its outer atmosphere), Eugene Parker of the University of Chicago developed a new theory about the Sun and its emissions. Logic suggested that the pressure and density of the corona should drop off considerably at great distances from the Sun. But observations and calculations showed that the corona seemed to escape and flow away from the Sun. In a famous paper published in 1958, Parker showed how the corona would not only expand but also speed up until it became supersonic. This "solar wind" would then spiral through the solar system, filling the space between the planets. Parker also showed how the solar wind would carry the solar magnetic field into interplanetary space. His idea that the solar wind could carry its own magnetic field would eventually become crucial to understanding how solar activity could affect the Earth and its field.

With the launch of the Sputniks by the Soviet Union and the Explorers by the United States during the first International Geophysical Year (1957–1958), centuries of scientific theories, remote observations, and wild speculation were finally confirmed or rejected by firsthand observation.

Equipped with a Geiger counter developed by James Van Allen, George Ludwig, and other scientists at the University of Iowa, the Explorer 1 satellite was launched on January 31, 1958. One of the goals of the flight—besides proving that the United States could orbit a satellite, as the Soviets had done with Sputnik 1 and 2—was to detect high-energy particles in space. Van Allen and his team were expecting to detect radiation from cosmic rays and other astrophysical sources, as researchers had detected with scientific balloon flights to Earth's upper atmosphere. And, in fact, their Explorer 1 Geiger counter did detect such rays. But during the flight, the high-energy particle counts rose higher than expected, reaching the top of the scale before dropping out completely. Since Explorer 1 did not have a data recorder, Van Allen's team could receive data only when the satellite passed over a few ground stations. Hence, they didn't know whether the loss of data was a glitch in the instrument, a new space phenomenon, or a problem with the radio signals.

When Explorer 3 was launched on March 26, 1958—this time with a data recorder—the mysterious readings of the Geiger counter became clear. As the satellite rose up to the apogee of its orbit, the particle counts rose steadily until they reached the highest level, stayed at the maximum for a while, and then abruptly dropped to zero. On the satellite's way back down to perigee, the reverse process occurred. Van Allen and colleague Carl McIlwain realized that the Geiger counter dropped to zero because the radiation in space had saturated their detector—there were too many particles for the instrument to count, so it shut down. Earth was surrounded by high-energy particles, prompting University of Iowa scientist Ernie Ray to exclaim: "My God, space is radioactive!" Analysis of the Explorer 1 and 3 data led Van Allen and his team to declare that two doughnut-shaped regions of magnetically trapped electrons and protons surrounded Earth. Colleagues named the feature the "Van Allen radiation belts."

Actually, the radiation belts could have been named the "Vernov belts" if Cold War politics had not gotten in the way of science. Sputnik 1 did not include any scientific instruments, but Sputnik 2 carried a simple set of spectrophotometers for detecting solar radiation, built by Russian scientist Sergei Vernov. As with the U.S. Explorer 1 satellite, Sputnik 2 could only relay information when the satellite was within radio visibility of the Soviet ground stations. So the data received were spotty and hard to interpret. Vernov's instrument had detected some increases in radiation levels in space, but there was no way to know if the phenomenon was local or global.

In May 1958, Sputnik 3 carried the first substantial package of scientific instruments on a Soviet flight, including another of Vernov's Geiger counters. To overcome the problem of radio contact from Russia, engineers built a tape recorder to collect and save the data for later transmission. The tape recorder did not function properly in ground tests, but to the bewilderment of scientists the head of the Soviet rocket program, Sergei Korolev, ordered the launch of Sputnik 3 before the problem could be fixed. As the scientists feared, the tape recorder did not work properly during the flight, and the scientific payoff of the mission was meager.

To make matters worse, Vernov's instrument did detect some tantalizing and puzzling increases in radiation during the real-time radio passes, but he was never able to retrieve the most useful data. The apogee of the orbit—when it would have passed through the belts—occurred while the satellite was over Australia. Scientists down under tracked Sputnik, but when they asked the Soviets for the key to the radio signals, they were rebuffed. Vernov's instrument probably collected enough data to fully map the inner and outer radiation belts, but instead that first scientific prize of the Space Age was left for Van Allen and his Explorer 4 mission in the summer of 1958.

Vernov's protégé, Konstantin Gringauz, ran into Korolev years later while walking in Moscow's Gorky Park. Gringauz asked Korolev why he had ordered the launch when everyone knew that the tape recorder would not work. Korolev told him that the order had come directly from Nikita Khrushchev. Thrilled by the successful launches of Sputnik 1 and 2, the premier had asked his scientists for another Soviet spectacle in time to boost the fortunes of the Italian Communist party in that country's elections.

In 1959, Gringauz and colleagues restored some Russian scientific pride when they used "ion traps" (plasma particle detectors) flying on the Soviet Lunik 2 and 3 spacecraft to detect the first traces of the solar wind flowing into the magnetosphere "cavity" around Earth. Three years later, when the U.S. Mariner II flew outside Earth's protective magnetosphere toward Venus, it detected a solar wind that flowed constantly. That solar wind fluctuated in fast and slow bursts and had periods of 27 days, as Maunder had suggested 60 years before. Gene Parker and other solar wind theorists were vindicated.

Finally, by the early 1970s, scientists were able to get above the murky atmosphere of Earth and take sophisticated telescopes into space for a better look at the Sun and the space around Earth. In 1971, a science team led by Richard Tousey of the U.S. Naval Research Laboratory in Washington, D.C., detected the first signs of Chapman and Ferraro's plasma clouds from the Sun with the NASA Orbiting Solar Observatory 7. Their coronagraph—which created a false eclipse by blocking the visible disk of the Sun from the field of view—allowed them to spy on the solar atmosphere. What they saw was a complex cluster of fast-moving clouds of plasma erupting from the Sun.

Three years later researchers from the High Altitude Observatory (HAO) in Boulder, Colorado, built a more sophisticated coronagraph that was sent up to the Skylab space station. Bob MacQueen and his science team at HAO were able to take time-lapse photos of these blobs of plasma, which would eventually come to be known as coronal mass ejections. Dave Rust and colleagues from American Science and Engineering used their own Skylab instrument—a soft X-ray telescope—to discover holes in the corona of the Sun, where Parker's solar wind could periodically stream toward Earth. Just as Maunder had suspected 100 years before, Rust and colleagues found that solar wind jetted out of holes in the corona and the streams would sweep past Earth in a cyclical pattern of 27 days (one solar rotation), much like the beacon from a lighthouse rotates its shaft of light in and out of view.

With large numbers of energetic particles trapped in the space around Earth and present throughout the solar system—and pictures of great blobs of plasma peeling off of the Sun—the puzzling circumstantial evidence that led Carrington, Loomis, and Stewart to connect solar activity with earthly fireworks was no longer a matter of speculation. It was now an observed fact.

The Cold War competition to explore space produced a few other surprises, not all of them pleasant. Suspicious of the possible source of the radiation in the Van Allen belts—nuclear weapons exploded by their enemies—and fearful of the effects of exploding weapons in space, the United States and the Soviet Union conducted several nuclear "tests" in the magnetosphere. Military leaders and scientists were partly motivated by physicist Nicholas Christofilos's prediction that nuclear explosions in near-Earth space could produce artificial radiation belts with significant military effects.

To test Christofilos's theory, the United States conducted Operation Argus in 1958, the "world's largest scientific experiment." From ships in the South Atlantic, about 1,000 miles southwest of South Africa, the Navy launched three modest warheads to 100, 182, and 466 miles in altitude. They studied how the energetic particles interacted with Earth's magnetic field, with radar and communications devices, and with the electronics of satellites and ballistic missiles. They found that Christofilos was right, as did the Soviets in similar experiments. Space radiation could indeed disrupt the work of man-made instruments and electronics.

The experimentation reached its absurd, *Dr. Strangelove*-style climax with the Starfish High Altitude Nuclear Test program in 1962. From an island near the equator in the middle of the Pacific Ocean, the United States launched a 1.4-megaton nuclear bomb about 300 miles into space. The explosion supercharged the Van Allen radiation belts and created artificial belts 100 to 1000 times stronger than normal space radiation levels. The high-energy electrons damaged the solar arrays of several satellites and caused three of them to fail. The electromagnetic pulse generated by the test led to power surges in electrical cables in Hawaii, blowing fuses, streetlights, and circuit breakers. Residual radiation from the experiment lingered in the magnetosphere for nearly seven years.

The Argus and Starfish tests were a stern warning to the nascent satellite industry and to military leaders that space might be a dangerous and difficult place to work.

5 Living in the Atmosphere of a Star

Deep beneath the surface of the Sun, enormous forces were gathering. At any moment, the energies of a million hydrogen bombs might burst forth in the awesome explosion. . . . Climbing at millions of miles an hour, an invisible fireball many times the size of Earth would leap from the Sun and head out across space.

Arthur C. Clarke, "The Wind from the Sun"

Space weather starts inside the Sun and ends in the circuits of man-made technologies. Defined simply, space weather is a range of disturbances that are born on the Sun, rush across interplanetary space into Earth's neighborhood, and disturb the environment around our planet and the various technologies—cell phones, satellites, electric power grids, radios—operating in that environment. The key to space weather is the transformation of energy, a transformation from magnetic energy and intense heat on the Sun to plasma energy in interplanetary space to magnetic and electrical energy around the Earth.

To the unaided human eye, the space between Sun and Earth appears to be a vast, dark void and the Sun is a static, unblemished fireball. But centuries of scientific observations and theories, as well as four decades of space exploration, have revealed that the space between our home and our star is windier than a mountain peak and as electric as a city night. And the space carved out by the Sun is hardly empty. It is filled with invisible electric and magnetic fields, high-energy particles, and a substance that's neither solid nor liquid nor gas. If a gas gets hot enough, the atoms—which are normally composed of a nucleus of positively charged protons and electrically neutral neutrons, surrounded by a cloud of negative electrons equal in number to the protons—start to lose electrons and become electrically charged. The electrons basically "boil" off the atoms, leaving a gas of free electrons and atomic nuclei that are positively charged. The result is an electrically conducting gas of disassociated electrons and nuclei called plasma—the fourth state of matter. Most of the universe, including our nearest star, is in the plasma state.

Gusty streams of this plasma continuously blow out from the Sun and rain down on Earth in a torrent of matter and energy; more energy escapes the Sun in some of its storms than humans have consumed in the entire history of civilization. In effect, we live inside the atmosphere of a stormy star, an atmosphere that stretches to the edge of the solar system and pushes against the interstellar plasmas of the Milky Way galaxy. That solar atmosphere has changeable weather and a dynamic climate, with the cosmic equivalent of winds, clouds, waves, hurricanes, and blizzards, all waxing and waning on cycles from minutes to months to millennia.

In astronomical terms, the Sun is a middle-sized, middle-aged gas ball that is just one of 100 billion stars in our Milky Way galaxy.[1] But our star is by far the largest object in our neighborhood, making up nearly 99 percent of the mass of the solar system (333,000 times the mass of the Earth). It would take more than

100 Earths to span the width of the Sun and more than a million Earths could fit inside. Even though the Sun is 93 million miles away—and its light takes 8 minutes, 20 seconds to reach Earth—it dominates life on our planet.

The main component of the Sun is the most basic chemical element, hydrogen, which is composed of one proton and one electron. It accounts for 92 percent of the Sun's atoms, though it is not quite precise to classify these most basic chemical constituents of a star as atoms. In the core of the Sun, the pressure is 250 billion times more intense than what we experience at the surface of the Earth, the density is 10 times the density of gold, and temperatures approach 16 million degrees Celsius (29 million degrees Fahrenheit). With conditions so extreme, the Sun's hydrogen atoms are actually broken up into their constituent parts—protons and free electrons. These particles are so closely packed together and moving so fast (the very definition of heat is mass in motion) that they collide frequently and combine in fusion reactions, producing helium and releasing energy.

This process of nuclear fusion actually goes through a number of complicated steps that took many years to figure out. The bottom line is that the amount of hydrogen mass going into the reaction is greater than the amount of helium mass coming out. Where does the missing mass go? It gets converted directly into energy. Every second, 700 million tons of the Sun's hydrogen is fused into 695 million tons of helium. About 5 million tons of mass is lost along the way, being converted to energy (principally gamma rays) that eventually reaches us as sunlight, solar wind flow, and other forms of solar radiation.

Currently, helium makes up about 7.8 percent of the Sun, a percentage that will grow over the next 5 billion years as the star slowly fuses all of the hydrogen. All the rest of the chemical elements, such as iron, carbon, and oxygen, make up less than 0.1 percent of the Sun, an astonishing figure when you consider that Earth and its atmosphere are made up almost entirely of those heavier elements. (Earth has very little hydrogen and helium in pure form. Almost all of our hydrogen is tied up in water; traces of helium are found in the atmosphere, while the rest is trapped principally in methane gas wells.)

Just above the core of the Sun, the "radiation zone" allows this nuclear reactor to move energy and light from the center toward the surface (see the color Plate 2 for a diagram of the Sun). Somewhat cooler than the core (about 5 million degrees C, 9 million degrees F), the radiation zone is comprised of plasma that passes on the energy produced below. Energy passes from particle to particle, gradually working toward the outer edge of the radiation zone. As Michelle Beauvais Larson of Montana State University describes it, the process is similar to standing in a crowded room where each person holds an empty glass. Imagine there is a sink at one end of the room and the people at the opposite end want a drink, but the room is so packed that no one can move. If the room were like the inside of the Sun, the person nearest the sink would fill his glass with water (energy) and pour it into the glasses next to him. The people with water in their glasses would do the same thing. This process could continue until the water is passed across the room. Energy moves from atom to atom in the radiation zone in a similar way, though not as neatly or systematically. In fact, it takes nearly 170,000 years for radiation to bounce its way from the Sun's core to the outer edge of the radiation zone (that figure is but one of many estimates of the time it takes for radiation to ping-pong its way out of the zone, as there are many different models of the activity inside the Sun).

Between the radiation zone and the visible surface of the Sun is the convection zone, where superheated gases rise from the interior like water boiling in a pot. Bubbles of hot plasma churn and circulate from the interior to the surface, release some of their energy, and descend back toward the radiation zone. The trip from the radiation zone to the photosphere—the yellowish-white sphere of light that we see—takes about one week. The Sun gives hints of this boiling action in the granulations or bubbles that percolate on the visible surface, where temperatures decline to about 6,000 degrees C (10,000 degrees F). The energy from the interior is released in the many forms of electromagnetic energy: X rays, ultraviolet and infrared light, gamma rays, radio waves, and visible light.

While plasma and energy swirl and circulate from the core to the photosphere, the Sun is also spinning its entire gargantuan mass around an invisible axis, just as the Earth spins. One of the most intriguing and important quirks of the Sun is its differential rotation; that is, the Sun rotates with different periods from the poles to the equator and from the surface to the interior. By charting and measuring the motion of sunspots across the visible face of the Sun, British scientist Richard Carrington and others determined in the latter half of the 1800s that on average the Sun makes a complete rotation once every 27 days. But in fact, the surface of the Sun at the equator makes a full rotation in as little as 25 days, while the surface at the North and South Poles can take as long as 36 days to spin once around.

Similarly, the various layers of the solar interior (core, radiation zone, and convection zone) seem to move at different relative speeds, and the entire outer layer of the Sun is slowly but steadily flowing from the equator to the poles. In 1997, scientists using instruments on the Solar and Heliospheric Observatory, a joint satellite mission of the European Space Agency (ESA) and NASA, found that "rivers" of hot plasma flow beneath the surface of the Sun. Like the trade winds on Earth, these rivers of plasma transport gas beneath the Sun's fiery surface. "We have detected motion similar to the weather patterns in the Earth's atmosphere," noted Jesper Schou, a solar scientist at Stanford University who was part of the team that discovered the flows. "And we have found a jet-like flow near the poles that is completely unexpected, and cannot be seen at the surface." These 40,000-mile-wide belts in the northern and southern hemispheres of the Sun flow at different speeds relative to each other, and they all move slightly faster than the solar material surrounding them. The belts reach at least 19,000 kilometers (12,000 miles) below the Sun's visible surface.

Like stripes on a barber's pole, these river-like bands start in the Sun's middle latitudes and gradually move toward the equator during the 11-year solar cycle, according to Craig DeForest, a solar physicist at the Southwest Research Institute. "They appear to have a relationship to sunspot formation as sunspots tend to form at the edges of these zones," DeForest says. "We speculate that the differences in speed of the plasma at the edge of these bands may be connected with the generation of the solar magnetic cycle which, in turn, generates periodic increases in solar activity." The result of this swirling mess—of matter and energy moving from the core to the surface, from east to west, all at different speeds—is that the Sun generates a complex global magnetic field that might explain everything from sunspots and coronal mass ejections to the solar cycle. Solar physicists call that process the solar dynamo.

In nature, electric currents produce magnetic fields, and changing magnetic fields produce electric currents. Within the Sun all of this flowing plasma generates electric currents, since plasma is a natural electrical conductor. Though they cannot yet explain how, scientists suspect that the shearing action between the moving plasmas on the edges of the radiation and convection zones produces intense electric currents that induce a global magnetic field throughout the Sun that stretches out into space. The magnetic fields are a bit like rubber bands, consisting of continuous loops of lines of force that have both tension and pressure. Like rubber bands, magnetic fields can be strengthened by stretching them, twisting them, and folding them back on themselves. The fluid-like flows of plasma at different speeds inside and at the surface of the Sun do just that—they wind the Sun's magnetic field into a tangled mess of loops and knots that poke out through the solar surface and stretch into its mysterious atmosphere (see Figure 8).

For the sake of space weather, the most important part of the Sun is its atmosphere, known as the corona. Starting just above the visible surface—the photosphere and the transitional "zone of color" or chromosphere—the corona stretches millions of miles away from the Sun. Typically, the corona is only visible from

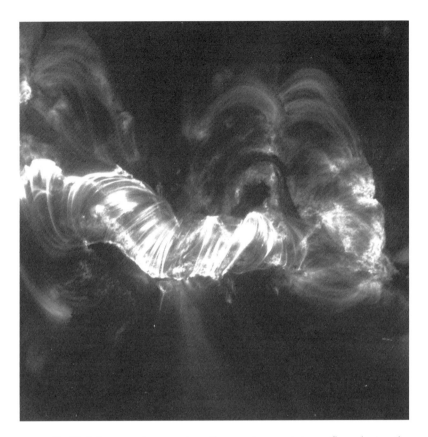

FIGURE 8. Tightly wound magnetic coils snap into an intense flare during the Bastille Day space weather event of 2000. NASA's Transition Region and Coronal Explorer (TRACE) spacecraft captured this close-up view. The area in these images is about 186,000 miles across, large enough to span 23 Earths. The Slinky-like formation of coronal loops was roiling at nearly 2.7 million degrees C. Courtesy of NASA.

Earth during eclipses, but the invention of the coronagraph and of space-based telescopes has allowed scientists to detect the faint light of the corona (about as bright as the full Moon). In those glimpses, they have spied magnetic loops and lines flowing with plasma and hanging above the surface of the Sun. They have also found gaping holes in the corona where the magnetic field lines from the Sun stretch out into space and high-speed solar wind gushes out.

The temperature of the Sun from core to surface makes a steep decline, such that the photosphere is about 6,000 degrees C. Despite its residence above the surface, the corona is actually hundreds of times hotter than the photosphere, reaching temperatures in the millions of degrees. How the corona can be so much hotter than the surface remains a grand mystery of solar physics. Most researchers surmise that it is related somehow to the dissipation of energy and the interaction of that energy with the complicated magnetic fields that burst from the interior and extend above the surface in great arches and loops. But no one in the field of solar physics has been able to offer an acceptable or understandable explanation because the causes of this unusual heating have been hard to detect. Deciphering the mystery of the heating of the corona is considered crucial to understanding why the Sun has weather.

Technically, the corona does not end: the high-pressure, million-degree plasma pushes the corona outward from the Sun to become the solar wind. Solar researchers have found that the electrified plasma of the solar wind flows out of the corona like water gushing through cracks in a dam. The solar wind essentially seeps out through the edges of honeycomb-shaped patterns in the surface of the Sun, escaping around the edges of large convection cells bubbling up from the interior. "If you think of these cells as paving stones in a patio, then the solar wind is breaking through like grass around the edges, concentrated in the corners where the paving stones meet," said Dr. Helen Mason, of the University of Cambridge, England, during a 1999 press conference. "However, at speeds starting at 20,000 miles per hour at the surface and accelerating to over 2 million miles per hour, the solar wind grows much faster than grass."

The expanding, speeding plasma of the solar wind races away from the Sun in all directions to fill the space between the planets. Each bubble of plasma rises from inside the Sun and carries an imprint of the magnetic field of the Sun embedded in a mix of ions and electrons and helium nuclei. Blowing at 800,000 to 5 million miles per hour, the solar wind carries 1 million tons of matter into space every second (that's the mass of Utah's Great Salt Lake poured out every second). Yet the solar wind would not even ruffle the hair on your head. Given the vastness of space, all of the mass is quickly spread out to a point where it has fewer particles per cubic centimeter than the best vacuums scientists can produce on Earth. Our own air is millions of times denser than the solar wind, such that 1 cubic centimeter of earthly air has as many particles as one cube of solar wind measuring 10 kilometers on a side.

This solar wind flows past Earth like water past a cruising boat. Tenuous compared to air, the solar wind is still potent enough to confine Earth's magnetic field, molding it into the shape of a comet or wind sock. Earth's magnetic cavity or cocoon keeps almost all of the Sun's harmful radiation and solar wind particles from reaching us (except for the ultraviolet rays that give you a sunburn). Less than 1 percent of the solar wind penetrates the magnetosphere, but that is enough to act as a cosmic generator, producing several million amps of electric current. This interaction supplies almost all of the energy in Earth's magnetosphere.

In addition to the flow of the plasma, the solar wind carries with it the solar magnetic field. The amount of electrical energy transferred from the solar wind to the Earth's magnetosphere depends on the north-south orientation of that solar wind magnetic field (known as the interplanetary magnetic field or IMF). If the magnetic field in the solar wind is directed southward, it can interconnect with the Earth's northward-oriented magnetic field. This direct magnetic connection allows energy to flow more freely between Sun and Earth, powering up the cosmic generator around our planet. Even though a mere fraction of the solar wind energy penetrates the magnetosphere under the worst of conditions, it's enough to cause global magnetic storms and auroras around Earth.

The most studied aspect of space weather is the sunspot, the first solar disturbance that was suspected to affect Earth. Appearing as dark patches against the bright background of the rest of the Sun, sunspots are actually relatively cooler than the surface (3,800 degrees C compared to surrounding temperatures of 6,000 degrees C). Discovered thousands of years ago by observers in ancient China, sunspots are regions of intense, complicated magnetic activity on the Sun, with magnetic fields 1,000 times the strength of Earth's field. Sunspots tend to appear in groups, and almost always in pairs of opposite magnetic polarity. They can last from several hours to several months, and they can be as large as 20 times the size of Earth (see Figure 9).

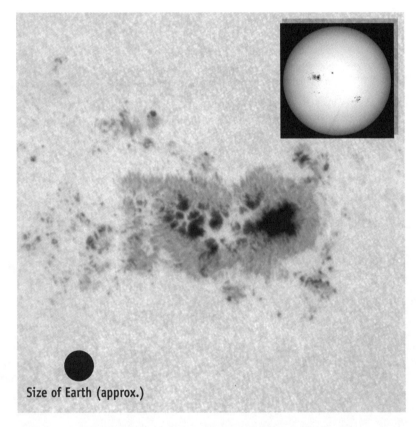

FIGURE 9. Caused by intense magnetic fields emerging from the interior of the Sun, a sunspot appears to be dark when contrasted against the rest of the solar surface because it is slightly cooler than the rest of the visible surface. Courtesy of SOHO/ESA and NASA.

Though sunspots have been considered the impetus for magnetic storms and other space weather effects around Earth, Space Age observations have shown that, as solar researcher Tom Bogdan of the High Altitude Observatory, Boulder, Colorado, puts it, "sunspots are more like symptoms than the disease itself." The active regions above sunspots (in the corona) emit X rays and radio waves that can hinder radio communication from Earth. But mostly, the spots are warning signals telling scientists that the Sun is more or less turbulent and disturbed and where the next solar blast might lift off. Over longer timescales, some researchers believe that sunspots may be an indicator of climate change on Earth. Scientists are still figuring out the role of sunspots in space weather, but they do know that when a flare erupts, sunspots are often nearby.

From 1859—when Carrington made the first observation—until well into the 1990s, flares were considered to be a principal driver of space weather effects at Earth. The strongest flares occur just several times per year, while weaker flares are relatively common, with as many as a dozen a day occurring during the Sun's most active periods. But as scientists have scrutinized these explosions, they have found that they are not necessarily the dominant factor in producing magnetic storms and auroras around Earth. That does not mean they are any less potent when it comes to raw power.

Solar flares appear as sudden, intense flashes in the chromosphere of the Sun. Flares occur when magnetic energy built up in the solar atmosphere is suddenly released in a burst equivalent to millions of hydrogen bombs. Scientists estimate that enough energy is released in one flare to power the United States for 20 years at its current level of consumption. On the other hand, the energy released is less than one-tenth of the total energy emitted by the Sun every second. The most commonly accepted model of solar flares suggests that the explosion creates high-energy electrons that funnel down toward the solar surface and produce X rays, microwaves, and a shock wave that heats the surface. The explosion also produces seismic waves in the Sun's interior that resemble earthquake waves. The seismic energy released at the surface is estimated to be 40,000 times the energy released in the great San Francisco earthquake of 1906.

Flares energize the particles in the corona—cooking them to tens of millions of degrees Celsius—and accelerate the particles in the outflowing solar wind to a point where radio waves, X rays, and gamma rays are shot out across the solar system, sometimes in the direction of Earth. Most of the particles are deflected by Earth's magnetic field and the atmosphere absorbs nearly all the harmful radiation, but flares still can have a crippling effect on space-based activities. The intense X rays from a solar flare travel to Earth at the speed of light, giving space weather watchers little time to react. A corresponding blast of high-energy particles (known as a solar proton event) reaches the magnetosphere in as little as 20 or 30 minutes. The onslaught of radiation heats the gases in Earth's upper atmosphere and causes the uppermost layers to swell, sometimes to a point where the increased friction can drag satellites down from their orbits prematurely. Long-distance radio signals can be disrupted and sometimes even blacked out by the resulting change in the Earth's ionosphere. The sensitive electronics and microchips of satellites can be pierced by the high-energy particles. Finally, the energetic particles accelerated in solar flares are dangerous to astronauts and, at times, even to occupants in some high-flying airplanes.

The most important solar event from Earth's perspective is the coronal mass ejection (CME), the solar equivalent of a hurricane. A CME is the eruption of a huge bubble of plasma from the Sun's outer atmosphere. Essentially, the corona rips open and blasts as much as 100 billion tons of material into space—equivalent to 100,000 battleships (but less than 46 quintillionths (45.6×10^{-17}) of the mass of the Sun). They are the largest structures that can erupt from the Sun and are one of the principal ways that the Sun ejects material and energy into the solar system. CMEs expand and fly away from the Sun at 1 million to 5 million miles per hour, traveling as huge magnetic clouds across the solar system—and sometimes through our neighborhood in space.

The buildup and interaction of magnetic loops hanging in the Sun's corona—which can stretch over, under, and around each other—seems to supply the energy to heat the corona and produce the violent explosion of a CME. Spiro Antiochos, a solar theorist at the U.S. Naval Research Laboratory in Washington, D.C., compares this process to that of filling helium balloons. If you inflate a balloon without holding it down, it will slowly drift upward. But if you hold the balloon down with a net, you can generate a lot of force when you fill it, causing it to push upward. Once you remove the net, the balloon shoots skyward. After observing how magnetic fields abut and interact, Antiochos and colleagues theorized that the Sun's magnetic fields intertwine and overlap like a net, restraining each other and forcing the buildup of tremendous energy. As the stressed field continues to emerge from the solar interior, it builds up more potential energy, pushes harder against these magnetic ropes, and moves higher into the corona. Eventually, through a process known as "magnetic reconnection"—in which opposing magnetic lines of force merge and cancel, releasing the stored magnetic energy—the field is released from its bonds and escapes the Sun at great speed.

Just hours after blowing into space, a CME cloud grows to dimensions exceeding those of the Sun itself, often as wide as 30 million miles across. As it ploughs into the solar wind, a CME can create a shock wave that accelerates particles to dangerously high energies. Behind that shock wave, the CME cloud flies through the solar system bombarding planets, asteroids, and any other object in its path. If a CME erupts on the side of the Sun facing Earth, and if our orbit intersects the path of that cloud, the results can be spectacular and sometimes hazardous.

Coronal mass ejections occur at a rate of a few times a week to several times per day, depending on how active the Sun may be. And because of the size of the plasma clouds they produce, the odds say Earth is going to get hit by a CME from time to time. Like flares, the fastest CMEs can accelerate particles in interplanetary space to the point where they can harm spacecraft or astronauts. CMEs also can produce shock waves and disturbances as they push into slower-moving solar wind, piling particles and energy in front of them like snow on the front of a plow.

As those particles reach Earth, the magnetosphere deflects most of them back into space. "It's like a never-ending football game," says Philippe Escoubet, project scientist for the European Space Agency's Cluster mission to study the magnetosphere. "The Sun is kicking particles like balls. The Earth is one of the goals and its magnetic field is the goalkeeper. The magnetosphere is always trying to push the balls away, but some get past. When particles score goals they disrupt the Earth." Much of the time, those energetic particles trickle in through the rear flanks of Earth's magnetic tail—on the nightside—and near the polar regions. Solar plasmas collect in the Earth's magnetic tail, mingling with earthly plasmas that have escaped our upper atmosphere. High-energy electrons and protons spiral along the planet's invisible magnetic field lines, suspended in space by magnetic and electric forces. They accumulate around the equator of Earth in the radiation belts and the tail of the magnetosphere in a dense region known as the plasma sheet.

However, it is not the particles from a CME that produce auroras and magnetic storms. The power of a CME lies in its ability to drive currents in Earth's magnetosphere—just as Chapman and Ferraro had proposed in 1930—and to energize the plasma that already surrounds the Earth. When a CME is directed at Earth and the conditions are just right, the ramming pressure of the cloud and its shock wave can make Earth's magnetosphere resonate like a bell struck by a hammer, exciting the electrons and ions trapped in the tail and radiation belts. Most importantly, if the magnetic field carried by the CME has a southward orientation (opposite Earth's northward-flowing magnetic field lines), the magnetosphere gets a major jolt. The magnetic field of the CME merges with the magnetic field on the dayside of Earth, transferring enormous amounts of energy to the magnetosphere in the process.

This interconnection, or reconnection, between the magnetic field of the Sun and the magnetic field of Earth is what gives a CME "geoeffectiveness," as scientists describe it, that is, a strong effect on Earth. Electrical current systems in space and in the ionosphere intensify, inciting the complex activity known as a magnetic storm. Much of the energy from the CME is pulled into the tail (nightside) of Earth's magnetosphere, where it is briefly stored before it destabilizes the system and gets shot down magnetic field lines toward our atmosphere, producing auroras and other phenomena in the ionosphere. In essence, a CME knocks hard on the Earth's front door but actually comes in through the back door.

This energy transfer can cause the radiation levels in near-Earth space to skyrocket. The high-energy ions and electrons of Earth's Van Allen radiation belts become more numerous and more energetic. In fact, the whole magnetosphere becomes a hotter place as the energy of the CME increases plasma temperatures. The pressure gradients between the energized plasmas drive millions of amperes of electric current through the magnetosphere, currents that we detect as a worldwide decrease in the strength of Earth's magnetic field (as measured at the equator). Some of the current is diverted along the Earth's magnetic field lines toward the upper atmosphere (particularly the ionosphere). The flow of this current causes the atmosphere to warm and expand, increasing the density of gases at high altitudes. During big storms the amount of power dissipated by these currents in the northern hemisphere alone can exceed the electrical power generating capacity of the United States.

Though magnetospheric physicists have long puzzled over the mechanism that accelerates the particles inside Earth's cavity, recent observations suggest that the Van Allen radiation belts act as a sort of cosmic particle accelerator. The two concentric rings of radiation have long been known to vary greatly but have often been represented for engineering purposes by average models (see Figure 10).

For decades, space physicists theorized that the Sun and its solar wind provided most of the high-energy particles found in Earth's radiation belts. But observations from scientific and mili-

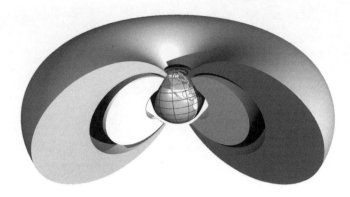

FIGURE 10. This cartoon of Earth's radiation belts reveals how one belt resides underneath/inside the other. The outer belt is largely made up of electrons and protons caught up in Earth's field by the interaction of solar wind and the magnetosphere. The inner belt is a product of cosmic radiation, which bombards the upper atmosphere and splashes energized particles into the space around the planet. Courtesy of Mike Henderson/Los Alamos National Laboratory.

tary satellites over the past 10 years show that the intensity of the belts can vary by 10, 100, or even 1,000 times in a matter of seconds to minutes. The radiation belts react to blasts from the Sun by boosting electrons to near light speed, to a state in which they are known as satellite-harming "killer electrons." "The radiation belts are almost never in equilibrium," says Geoff Reeves of Los Alamos National Laboratory. "We don't really understand the process, but we do know that things are changing constantly."

According to Reeves, there is no way that the solar wind or a CME alone could cause such a fluctuation in the particles trapped around Earth. "There are just not enough high-energy electrons in the solar wind to explain how many we observe near Earth," said Reeves. Dan Baker of the University of Colorado adds, "It's amazing that the system can take the chaotic energy of the solar wind and utilize it so quickly and coherently. We had thought the radiation belts were a slow, lumbering feature of Earth, but in fact they can change on a knife's edge."

Though scientists have not yet worked out the physics of the great particle accelerator in our sky, they believe the magnetosphere is an effective and efficient accelerator of particles that dwarfs any man-made supercollider or Tokamak.

There's no need to run for cover from space weather. Storms from the Sun cannot harm life on the surface of the Earth. But they do affect the way we live. With the average CME dumping about 1,500 gigawatts of electricity into the upper atmosphere, big changes occur in the space around Earth and the upper atmosphere. Those changes can suddenly upset the daily commerce of a world that has come to depend on satellites, electric power, and radio communication—all of which are affected by electric and magnetic forces.

As the next five chapters will show, each of those effects has already made front-page news on Earth, and they are likely to have an impact on our efforts to explore worlds beyond our own. The magnetic storms caused by all of these swirling particles and electrical currents from space can distort the magnetic field of Earth enough to wreak havoc on electrical power and ground-based communications systems (Chapters 6 and 7). X rays and high-energy protons from solar blasts can completely wash out radio communications for hours to days and can disrupt signals sent to and from satellites (more about this in Chapters 7 and 8). Super-energized particles from the radiation belts and from auroral storms can damage the sensitive electronics of satellites (Chapter 8). They can even harm an unprotected astronaut working in space and occasionally penetrate far enough into the atmosphere to give a mild dose of radiation to passengers on high-flying jets (Chapter 9). And in the most controversial and least understood effect of space weather, the long-term variation of solar activity could influence the climate patterns of Earth on scales from decades to millions of years (Chapter 10).

6 The Cosmic Wake-Up Call

I should study Nature's laws in all their crossings and unions;
I should follow magnetic streams to their source and follow the shores
of our magnetic oceans. I should go among the rays of the aurora,
and follow them to their beginnings, and study their dealings and
communions with other powers and expressions of matter.

John Muir

"It was the biggest thing any of us had seen," said scientist JoAnn Joselyn. "Before it was even visible you could see a depression on the limb of the Sun. It sent a chill down your back. It was just plain ugly."

"It" was sunspot region 5395. On March 4, 1989, observers at the U.S. Air Force's Ramey Solar Observatory in Puerto Rico detected a surge of activity just over the Sun's eastern edge, or limb. X-class flares, the most intense the Sun can produce, were shooting off into space from a region that scientists could not yet see. But what they could see, as Joselyn noted, was a notch in the side of the Sun where a sunspot was compressing the surface. When the gnarled patch of black spots rotated into full view early on March 6, observers got a view of the most complex sunspot region—both magnetically and structurally—that any of them had ever seen. More than 43,000 miles across, the sunspot group was 54 times the size of the Earth. It was pulsing and roiling at latitude 34 degrees, an unusual place for sunspots in the middle of solar maximum.

"There had never been a sunspot region like that in our lifetime," said Joselyn, a space weather researcher and onetime forecaster at the National Oceanic and Atmospheric Administration's (NOAA) Space Environment Center (SEC). As it came around to the front of the Sun and started its march across the face "we were just waiting for something big to happen," Joselyn said. "We were dumbfounded, and we felt a little helpless. Something was going to happen that we had never seen before."

From March 6 to 19, sunspot region 5395 exploded with at least 195 solar flares, 11 of them of the most intense "X-class" variety, and another 48 of the next highest "M-class" of severe flares (see Figure 11). The telescope on NASA's Solar Maximum Mission (SMM) satellite detected 36 coronal mass ejections over the course of those two weeks.

The opening salvo came early on March 6, when region 5395 flashed with one of the three most powerful flares ever observed.[1] The stream of radiation from the explosion lasted 10 hours—the norm is perhaps 30 minutes. Solar physicists estimated that the temperature inside the flare reached 20 million degrees Celsius, and more energy was released in those moments than humans have consumed in the entire history of civilization. The radio signals emitted by the flare were 2,000 times more intense than the normal background noise of the Sun. X rays from the explosion arrived at Earth in 8 minutes, 20 seconds (the time it takes light to travel from Sun to Earth), and protons from the Sun swept by a few minutes later like a swarm of angry hornets. The X-ray and proton detectors on the U.S. Geostationary Orbiting Environmental Satellite 7 (GOES-7) were flooded, as readings went off the normal scales for 27 minutes. It was labeled an X-15 flare.

Equipped with satellite data, ground-based telescope images of the Sun, and computer models, the scientists and forecasters at

The Cosmic Wake-Up Call

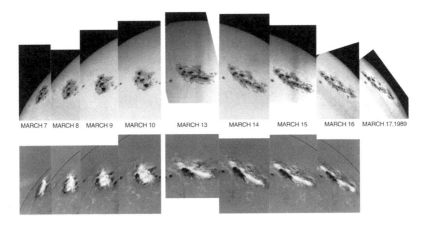

FIGURE 11. This time series of images reveals the evolution of sunspot group 5395 as it moved around the Sun in March 1989. The upper sequence, taken in white light, shows the usual representation of the solar surface. The lower images are magnetograms, where black and white regions show opposite magnetic polarity. Note that changes in the white-light images coincide with changes in the magnetic structures. Courtesy of NOAA/AURA/NSF.

SEC watched the biggest space weather event of their lives begin to unfold in real time. All six phones in the SEC forecast center were ringing constantly. Prophetically, lead SEC forecaster Joe Hirman told *The Washington Post* in a March 6 interview: "There is definitely a chance of more flares because the chances are that this spot will keep doing something."

The Sun continued to seethe and snarl with smaller flares for three days, and Earth's magnetosphere was bathed in a shower of high-energy protons. That stream intensified on March 9, when sunspot region 5935 unleashed an explosive flare that reached the peak of brightness (brilliance) and size on the scales solar physicists had devised. The next day observers witnessed a rare white-light flare, much like the one Carrington had seen 130 years before. The SMM satellite's coronagraph/polarimeter detected a large halo coronal mass ejection; the halo meant that the cloud was directed at Earth. The SEC predicted that there would be "high activity" on March 12 and 13.

Late on March 12, the biggest blast of energetic particles from the Sun started to reach Earth, as the rain of high-energy protons increased to 100 times the norm. By the middle of March 13, the magnetosphere, which normally stretches out 34,000 miles toward the Sun, was pushed down inside geosynchronous orbit (altitude 22,000 miles). Geosynchronous satellites that usually fly inside the protective cocoon of Earth's magnetic field suddenly found themselves passing out into the solar wind, as the magnetosphere shrunk to half its normal size (some estimate as low as 14,000 miles).

Around the world, sensors in magnetic observatories went off the top of their scales for five or six hours. The auroral electrojet—an electric current system swirling in the ionosphere about 60 miles above Earth—fluctuated wildly as it slid down the globe from its normal northern latitudes and flowed right over the heartland of the United States. U.S. Navy observers in Bay Saint Louis, Mississippi, watched their magnetometer get pinned to the top of its scale as auroras danced in the skies.

Vivid red auroral rays glowed over all over North America, stretching as far south as Arizona, southern California, Mississippi, and Texas. Police stations were flooded with phone calls about "funny red clouds" in the night skies. A group of backpackers in the hills of North Carolina recalled seeing the red glow in the sky and thinking that an extremely large fire or perhaps even a nuclear war had started somewhere over the horizon. Across the Atlantic, reports of aurora came in from brightly lit cities in Portugal, Spain, The Netherlands, Britain, and Hungary. In the southern hemisphere, sky watchers saw the *aurora australis* over New Zealand, Australia, and South Africa. The light show stretched to the tropics, where mystified observers watched the aurora in Cancún, Mexico, Grand Cayman Island, Honduras, and Dominica. One observer in Cuba said: "It was like night turning into day."

The *aurora borealis* was big news in Florida. A newspaper reporter from Miami described how "an eerie red-orange glow danced across South Florida's skies in what may have been a performance of the northern lights . . . [an occurrence] rarer than snow." Tom Printy of the Central Florida Astronomical Society told one reporter that "many of those who saw an aurora for the first time were baffled, thinking that the sky activity might be related to the space shuttle that was launched that morning."

That 13-day surge of solar activity caused one of the most powerful magnetic storms ever recorded.² In the two weeks that sunspot region 5935 marched across the face of the Sun, the forecast group at SEC issued 37 rapid alerts—urgent warnings of threats to technological systems on Earth—and 415 normal alerts. The words of caution helped, but not every industry has a practical solution for getting out of the path of a space weather storm.

In the northeastern United States, a manufacturer of computer microchips shut down operations because the magnetic storm was disturbing sensitive instruments. All around the world, navigators and surveyors noted that their compasses were distorted by as much as 10 degrees. Several North Sea oil companies stopped drilling because magnetic instruments that guide the drills were way off course, and most magnetic surveys and oil-prospecting missions came to a halt for hours to days.

Up in the thick of the storm, satellites felt the full force of the Sun. Heated and electrified by the solar protons and the compression of the magnetosphere, the density of the upper atmosphere of Earth increased by five to nine times the norm, increasing the friction—atmospheric drag—on satellites in low-Earth orbit. The U.S. Air Force Space Command at the Cheyenne Mountain Operations Center typically tracks 8,000 objects in orbit around the Earth—everything from bus-sized satellites to space junk such as floating nuts and bolts. By March 14, 1989, the Air Force had lost track of 1,300 of those objects, many of which had dropped by more than 1 kilometer in their orbits. NASA's SMM satellite was said to have "hit a brick wall," dropping nearly a $1/2$ mile in one day and a total of 3 miles from its normal orbit due to the increased drag. A classified U.S. military satellite in low-altitude orbit began an uncontrolled tumble through space.

Aside from the drag, some satellites took a more direct hit from the solar radiation. Bombarded by radiation, NOAA's GOES-7 weather satellite suffered communications circuit problems and outages for much of March 13, and Japan's CS-3B communications satellite lost half of its computer brain. Many satellites endured phantom switching and tripping of circuits in the electronic controls, and seven commercial satellites in geostationary orbit had trouble staying in place. Operators of the seven satellites fired thrusters at least 177 times in two days to keep their spacecraft on track, more adjustments than are typically made in an entire year. Other satellites had communications problems when the excited ionosphere interfered with radio signals sent to and from the ground. In a few instances, signals from the Global Positioning System were severely degraded or gave outright erroneous positioning information.

The most spectacular effect of the March 1989 storm—and the effect that has many scientists and engineers concerned as we live through the current solar maximum—was the burnout and blackout of several electric power systems around the world. Stray electric currents from the storm—known as geomagnetically induced currents (GICs)—disturbed at least 200 different components of electric power systems in Maryland, California, New York, New Mexico, Arizona, and Pennsylvania.

In Sweden, five power transmission lines were tripped by large voltage fluctuations, and fire alarms were set off. At another Swedish power company, Sydkraft, operators detected a 5 degree C increase in the temperature of the rotors of a nuclear plant. At Public Service Electric & Gas's nuclear power plant in Salem, New Jersey, a $10 million transformer was damaged beyond repair (see Figure 12). The extra current coursing through the system heated thick metal coils—bathed in cooling mineral oil—until they burned up. Typically, 500-kilovolt transformers like the one in Salem can take a year to replace, but PSE&G was fortunate enough to find an

FIGURE 12. Charred wires reveal the damage wrought by the March 1989 magnetic storm. This section was part of a large transformer at a nuclear power plant in New Jersey. Such hardware can cost tens of millions of dollars to replace, and it can take as long as a year to have a new one manufactured. Courtesy of John Kappenman.

unused spare from another canceled nuclear plant. The new transformer and plant were back in service within six weeks, but not before losing $400,000 per day in power sales.

Five hundred miles to the north, the Hydro-Quebec power company was not so fortunate. The Canadian utility suffered through a chain reaction that collapsed an entire power grid in 90 seconds. There was hardly time for the operators to assess what was happening, no less fix it. The press release from Hydro-Quebec's managers asserted that, "the March 13 blackout was caused by the strongest magnetic storm ever recorded since the 735 kilovolt power system was commissioned." At 2:45 a.m. the storm tripped five power lines from James Bay and caused a loss of 9,450 megawatts of power. "With a load of some 21,350 megawatts at that moment, the system was unable to withstand this sudden loss and collapsed within seconds."

More than 6 million people in Quebec City, Montreal, and surrounding areas lost power in the middle of a frigid winter night. The morning rush hour was snarled as traffic lights went out and the subway system shut down. It took nine hours to restore power to most of the region—through purchases of power from other utilities. But about 17 percent of customers went without power for much of the rest of the day, some for several days.

The cascade of problems began in the James Bay region shortly after the onset of the magnetic storm. In a total of 59 seconds, seven voltage-regulating devices at the Chibougamou, Albanel, Nemiscau, and La Verendrye substations tripped and stopped the flow of electricity. Voltages in the lines fluctuated by as much as 15 percent. With the sudden loss of voltage, all five major transmission lines to Montreal tripped, some of them exploding into flames. The loss of 9,450 MW of power overloaded the rest of the system. Two transformers blew out in Chibougamou, and the Churchill Falls and Manicouagan-Outardes power plants automatically shut themselves down as demand for power overwhelmed the system. Transmission lines from Sherbrooke failed, cutting off the export of energy to New England.

The Hydro-Quebec blackout resulted in a loss of some 19,400 MW of power in Quebec and 1,325 MW of exports to New England and other parts of Canada. The restoration of power took nine hours because much of the essential equipment, particularly on the James Bay transmission network, was made unavailable by the blackout. Operators had to use power from isolated stations that normally export their electricity or buy it from other power companies in Ontario and New Brunswick.

By the time the blackout was over and the fried equipment was repaired or replaced, Hydro-Quebec had lost at least $10 million. The cost to its customers was estimated to be in the hundreds of millions of dollars. In the months following the event, the Northeast Power Coordinating Council and Mid-Atlantic Area Council power pools—electric power cooperatives that serve the northeastern United States from New England to Washington, D.C.— acknowledged that they nearly suffered a cascading system collapse due to the stress of the storm and the loss of Hydro-Quebec's power exports. The whole northeastern United States almost tumbled into a blackout.

"The 1989 event is the most significant space weather event for the power industry," noted David Boteler, a researcher and electrical engineer who studies the effect of geomagnetic storms on power systems for the Canadian Geological Survey. "It was the event that changed people's opinion from 'space weather effects are just an academic curiosity' to 'this is a real problem that needs to be looked into.' Before 1989, believing in space weather effects on power systems was regarded by some as equivalent to believing in little green men from outer space."[3]

In high school physics, you learn that the way to generate electric current is to vary the magnetic field around a conducting wire. Move a bar magnet back and forth near the wire, as scientist Michael Faraday once proved, and the ammeter will show electricity flowing. That's essentially what happens when a power system is hit by a magnetic storm. From the perspective of space, electric power lines, railroad tracks, oil pipelines, and communications cables look a lot like long, thin, conducting wires. When space weather stirs up the great electric current in the sky—the auroral electrojet—the magnetic field at the surface of the Earth starts to vary. The result is a surge of extra electric current into power systems and every other sort of cable.

"Magnetic disturbances are causing induced currents to flow through the conducting networks that mankind has stretched across the Earth's surface during the last 150 years," David Boteler says. As electric potentials on the ground reach 1 to 10 volts per kilometer, the currents flow along Earth's surface. In the case of electric power systems in North America, direct current (DC) is pumped into lines that typically carry alternating current (AC). The DC saturates one-half of the AC cycle, causing power lines to be overloaded and transformers to try to compensate. This leads to a chain reaction. Regulating devices in some transformers may cause that part of the system to shut itself down. The rest of the system has trouble synchronizing and regulating the flow of energy, causing other devices and transformers to shut down in a spiraling effect. According to electrical engineer John Kappenman: "You just don't know what protective systems are going to shut off when you need them most."

The problem is exacerbated in Canada and the United States. "North America is the most profoundly affected land mass in the world because of igneous rock geologies which cover large regions," says Kappenman (who proudly notes that the power systems in Minnesota, which he helped oversee in 1989, did not fail on his watch). Igneous, or volcanic, rock tends to resist the flow of electric current more than other types of rock, forcing currents to flow closer to or at the Earth's surface. When a magnetic storm begins, GICs enter and exit power systems through ground wires attached to transformers, which would normally dump some of that current into the ground. Compared to igneous rock, however, a power line is the path of least resistance. Since the crust of Earth beneath North America is rich in igneous rock, and since the continent has so much infrastructure located at high magnetic latitudes (the magnetic north pole tilts toward Greenland and Canada), North American power systems are much more susceptible to GICs.

Statistical research confirms the geophysics. Kappenman and other researchers at Minnesota Power and Electric analyzed the failure rate of transformers in the United States from 1968 to 1991. They found that certain types of transformers failed much more often in GIC-susceptible regions; in fact, failure rates in the northeastern United States were 60 percent higher than in the rest of the country. The researchers also found that transformers had shorter life spans in those GIC-prone areas. Perhaps most compelling was the finding that transformer failures followed a periodic cycle that, as Kappenman notes, "virtually mimics" the solar cycle.

○●○

In 1990 researchers at the Oak Ridge National Laboratory studied a hypothetical space weather event just slightly more severe than the March 1989 storm. In their imaginary event, engineer Paul Barnes and economist James Van Dyke wiped out most of the tightly linked electric power grid from the Middle Atlantic up to New England in order to assess the economic impact. Their hypothetical magnetic storm began by sweeping across Canada and the United States at a time of day "when power import levels to New England and New York are near their limits and the capacity margins are at low levels." First, a Canadian utility cut off its exports to stabilize its own network. Then several voltage regulators were tripped and a few transformers were damaged. The extra load on the system and the loss of voltage in some of the transmission lines caused the "tie lines" from the Midwest and South to become overloaded and unable to supply the extra power needed for the Northeast. The cascade of problems flooded and wiped out large portions of the power grid. In the simulation it took 16 hours to restore half of the power and 48 hours to get the whole system back online.

Barnes and Van Dyke tallied the damage—the cost of purchasing power from outside utilities to cover the lost power generation, the cost of replacing fried equipment, and the revenues lost from the loss of power sales. They did not even begin to assess the economic losses to the local economies that would be halted by the blackout. Their total cost: $3 billion to $6 billion. The assessment by Barnes and Van Dyke, coupled with reviews conducted by the North American Electric Reliability Council, rated the hypothetical 1989 storm as an electric power disaster comparable to the damage caused by Hurricane Hugo.

That scenario, according to Kappenman, is not as improbable as it seems. Since 1965, electric power consumption in the United

States—particularly in the Northeast and California—has steadily increased, while very few power plants have been added to the infrastructure because of environmental protests and economic concerns. The result is that power grids are running so close to the operating margin that electricity is regularly being imported from Canada and the Midwest over "transmission lines operating near their limits." With less spare capacity on the lines, the risk of saturating the system during a magnetic storm rises substantially. In addition, most power companies across the continent have computerized their systems and linked them together to share power. Such power sharing helps utilities ensure that an isolated event does not bring down an entire power grid. But that also means there is a wider network that can get pulled down when faced with a widespread problem such as a global geomagnetic storm.

"The industry is more focused on making money than on protecting power," David Boteler notes, "more focused on accounting than engineering. The power-sharing strategy breaks down when there is a large event. By being interconnected, the companies are effectively making even longer transmission lines through which current can flow. Instead of 100 extra volts across 100 miles, you get 1,000 volts across 1,000 miles."

It would take billions of dollars "to plug the GIC sieve that the network has become," John Kappenman notes. The cost of rebuilding the existing power systems or retrofitting with devices that could block GICs is too prohibitive when weighed against the perceived risk. So Kappenman and several colleagues have been promoting specialized forecasting as the solution. With 30 to 45 minutes of advance warning, he notes, power companies could reconfigure the flow of power or shore up certain parts of the system to withstand the onset of a magnetic storm in the same way that they brace for a hurricane.

Monitoring conditions from the ground does not help, because by the time the ground-based observatories can detect magnetic fluctuations, the storm has already arrived. But by monitoring conditions in space on the sunward side of Earth, and by modeling what the varying solar wind will do to the magnetic field, power grid operators can steal 30 minutes of preparation time, Kappenman says. While he has been developing a computer model and software to provide such a forecast of ground-induced currents, Kappenman also has been lobbying industry and the government for years to invest in a reliable solar wind monitor to provide the data the power industry needs. In 1998, Kappenman's wish was fulfilled. With the launch of NASA's Advanced Composition Explorer (ACE) research satellite, Kappenman got the real-time data he needs to run his GIC modeling system. "Powercast," as he calls it, predicts when space weather might become dangerous to a power grid. That system—the world's first space weather prediction system for national electrical power grids—was put into operation in England and Wales in January 2000. During testing, according to Kappenman, the system was 95 percent accurate in alerting users to potentially hazardous current-inducing geomagnetic storms. Time will tell if it works in the real, sunny world.

But his efforts and those of several colleagues are being met with mixed interest. "Very few people have a real understanding of what GICs will do to a transformer," Kappenman says. "There are subtleties that no one fully appreciates." Furthermore, the life cycle of an electrical engineer working in system operations is just a few years, so there are not many people monitoring the power lines who remember 1989.

"An event like 1989 is not going to happen very often, and there is a danger that the risks get overstated," Boteler notes. "There are a lot of other things that can go wrong in a power system. And shutting down exports or turning on reserve power stations to prepare for a predicted magnetic storm can cost a lot of money. The question is whether you design power systems for the 'one-hundred-year event,' as you do for earthquakes and floods," Boteler adds. "It is amazing how often these 'one-hundred-year events' come around. Do you buy insurance, or do you hope it is not going to happen?"

For Kappenman and other colleagues now selling their own brand of electric power insurance, another 1989-style blackout could be very good for business.

7   *Fire in the Sky*

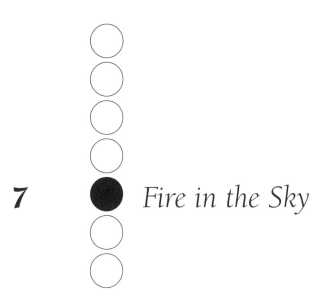

> The storm, when it struck, was a classic example of perhaps the oddest sort of foul weather the Earth is plagued with . . .
>
> John Brooks, "The Subtle Storm"

On the night of February 10, 1958, the brightest lights in New York City were not on Broadway. Instead, they were dancing in the sky over Central Park and the Battery. Just days after a brutal winter storm dumped snow and freezing rain on much of the East Coast (as far south as the Gulf states) and kept temperatures hovering between 0 and 20 degrees Fahrenheit, auroral rays and arcs were visible through the smoke and city lights of Manhattan. One of the most intense auroral displays over the Americas in the twentieth century occurred that night. Nature was painting the town, the snow, and the sky red.

The intensity of the great aurora and magnetic storm of February 1958 was nearly as big a surprise to some scientists as it was to the awed and startled public. Two days earlier, observers at the Sacramento Peak Solar Observatory in New Mexico detected a series of solar rumbles in the vicinity of sunspot region ABOO, a dark blotch that covered 3 billion square miles of the Sun's face. "Like the milder rumblings before a thunderclap, seven smaller flares had been counted that day before the big one came," journalist John Brooks wrote in an article for *The New Yorker*. The "big one" arrived at 2:08 p.m. Mountain Time on February 9, when a white-light flare burst into view. Four minutes after the onset of the big flare, the Harvard Radio Astronomy Station at Fort Davis, Texas, began hearing radio noise from the Sun. Brooks noted that during the greatest solar flares "the Sun sends out bursts of radio noise that, when picked up by special high-frequency receivers on Earth, sound like sausages being fried." The flashing and popping on the Sun lasted for nearly two hours.

News of this "whopping big" flare, as the staff of Sacramento Peak called it in their initial reports, was transmitted to Walter Orr Roberts at the World Data Center on Solar Activity at the High Altitude Observatory in Boulder, Colorado. As one of the lead U.S. scientists for the International Geophysical Year (IGY),[1] Roberts was charged with deciding whether or not to send an official advisory for a "special world interval." Such alerts—part of an international science program to make coordinated studies of geophysical phenomena—were intended to provoke scientists to look out for anything out of the ordinary. Despite the exuberance of the messages from the solar observatories, few people in the laboratories and agencies charged with monitoring space weather expected the storm to have such impact. Radio communication experts predicted that conditions would remain "fair to good." Noting that the flare was not necessarily the largest type (observations from other optical and radio observatories participating in the IGY did not agree with the "whopping big" classification from Sacramento Peak) and that "special world intervals" were a strain on limited science budgets, Roberts decided against rallying the troops. It was a decision that he would later openly regret. Within 28 hours, one of the biggest magnetic storms on record commenced.

The magnetic storm started around 8:30 p.m. Eastern Time on February 10, as magnetometer needles began to quiver at the World Warning Agency in Fredericksburg, Virginia, and in other stations around the world. Auroras sprang up in their usual northern regions but then began to descend toward the middle latitudes. "At one minute before nine, New York time, all magnetic hell broke loose from east to west and from Pole to Pole," wrote Brooks. A full magnetic storm commenced—a storm that is still ranked in the top 12 of recorded history.

"If the human organs of balance and orientation depended upon magnetism rather than gravity," wrote Walter Sullivan, the dean of modern science writers, in his book *Assault on the Unknown*, "every man in the world would have been dizzy, yet a large portion of this planet's population sat or slept comfortably at home, unaware of what was going on."

Those who were awake and attentive spied one of the most spectacular auroral light shows in U.S. history. From Canada to the Gulf of Mexico, rich red auroral lights (the calling card of the greatest magnetic storms) filled the heavens. That night the skies over almost all of North America were clear and the auroral displays were disproportionately concentrated over the continent. Reports of aurora sightings poured in from the usual places but also from Tulsa, Los Angeles, Havana, and Miami and from a steamship off the coast of Acapulco. The flickering, blood-red skies led to hundreds of false reports of fires. Sky watchers across the lower 48 states of America were treated to occasional turns of green and white auroras—typically visible only at high latitudes—and of auroral coronas—rays converging to a point, appearing like a shower of light. *The New York Times* noted that it was "one of the few occasions in which the aurora drapery had been seen here."

But all was not so pretty and bright for those working in the electric power and communications industries. As the magnetic storm raged through the night, huge geomagnetically induced currents surged through the wires and cables. In Ontario, circuit breakers were tripped by the storm, plunging Toronto into a short-lived blackout. Lights flickered in Minnesota, the Dakotas,

Montana, and British Columbia. Telephone and Teletype circuits were disrupted between the United States and Europe for nearly three hours due to excessive currents on the coaxial cable links from Newfoundland to Scotland. Telegram messages on the Western Union cables were garbled, as stray electric currents flowed from west to east through the lines and potential drops varied by as much as 320 volts. Telephone calls from the United States to Europe were received in alternating sequences of whispers and squawks.

Radio communication was not much better. From 9 p.m. to 11 p.m. on the East Coast, radio operators for American Telephone and Telegraph, *The New York Times,* and RCA Communications struggled to sustain their transatlantic radio signals. They rerouted their messages through radio stations at lower and lower latitudes until eventually those signals faded as badly as the west-east signals. In Boston, television viewers noted that two of their shows mysteriously swapped channels. By 11 p.m., all radio contact between the United States and Europe faded into silence for two to three hours. As Brooks wrote: "The Old World and the New were in scarcely better touch than they had been in the days of the clipper ships."

The radio blackout was particularly troublesome for the airline industry. More than 100 planes were "groping their way in one direction or the other" between Europe and North America during the magnetic storm. Under normal circumstances, they would have relied on radio transmissions to relay information on weather, traffic, and landing conditions. But on February 10, pilots found that they could only make radio contact if they were within a visible line of sight of a station. The cockpit airwaves were buzzing more than usual as pilots found themselves relaying messages from plane to plane. At least one pilot had to go it alone: an Air Force plane loaded with passengers and flying from New Zealand to Antarctica made the 2,000-mile journey over ice and frigid water without radio contact from anyone.

For some scientists the storm turned into a bonanza in spite of the fact that the event was not "official." Knowing that the solar storm was likely to stir up auroras and some fine scientific observations, space physicist John Winckler and a team of colleagues from the University of Minnesota braved a frigid night in order to launch a research balloon. A year earlier, Winckler's group had made perhaps the first "discovery" of the International Geophysical Year during an auroral storm. Launching an instrument-laden balloon within hours of the start of the IGY (which began at midnight, Greenwich/Universal time on July 1, 1957), the Minnesota scientists had detected X rays that seemed to emanate from the aurora. They surmised that energy had been absorbed from a solar storm and was transformed by auroral processes in Earth's ionosphere to be emitted as X rays. By February 1958, Winckler and colleagues were prepared to follow up their X-ray observations and do the sort of coordinated science that the IGY was designed to promote. Using fresh observations of auroral X rays from the night of February 10 to 11, they compared their data with those from cosmic-ray detectors, as well as radio and magnetic observatories on the ground. The researchers also compiled a synopsis of the space weather storm, detailing the flow of energy and activity from Sun to Earth.[2] They estimated that the cloud of solar material that had passed over Earth was 23 million miles wide and 46 millions miles long (half the distance to the Sun).

A century after Carrington and Loomis dissected the great storm of 1859, the link between solar storms, magnetic storms, and man-made technology was obvious, if not well understood. Wires invented to carry the electromagnetic currents of telegraph signals, telephone messages, and electricity—even streetcar and electric train equipment[3]—also picked up currents from the great generator in the sky (we now know them as geomagnetically induced currents, as noted in Chapter 6). Radio and television signals, as well as radar pulses, could sometimes be disrupted by the manic Sun. And as Brooks speculated in his story about the 1958 magnetic storm, "Nobody knows what kinds of apparatus still undreamed of may come along to be thrown out of whack by their caprices."

"The recording of this event on a global scale was one of the major achievements of the IGY," wrote Walter Sullivan. "As with the thorough examination of a patient—making use of electrocardiograms, a half-dozen laboratory tests, and direct observation—the mass of assembled information may make it possible to diagnose more accurately the exact nature of a magnetic storm."

One of the most ubiquitous and useful technological tools of the 1950s was the radio wave. Discovered at the end of the nineteenth century and harnessed in the first half of the twentieth, radio waves were used for wireless transmission of music, news, and drama to the public; for ship-to-shore, air-to-ground, and other navigation and transportation-related communications; for the detection of incoming planes and missiles in wartime (via radar); and for transmission of the burgeoning entertainment form of the era, the television program. Radio waves made the world smaller, safer, better informed, and somewhat more entertaining. They also brought scientists and engineers face to face with another aspect of space weather.

At the upper edge of the atmosphere, where Earth's environment meets the space environment, solar radiation breaks gases into the ions and electrons of plasma, forming a region called the ionosphere. This plasma-filled ionosphere conducts electric current and reflects radio waves, making most modern communications possible. Radio signals bounce off of the seemingly "flat" surface of the ionosphere as if it were a mirror reflecting light, and the behavior of these transmissions can be predicted. This phenomenon allows radio operators to overcome the curvature of Earth and transmit signals over the horizon, bouncing signals off the edge of space to reach distant receivers.

During space weather storms, however, the density of plasma in the ionosphere can be quite variable, becoming agitated by various environmental changes and forming "clumps" of plasma. Disturbed patches of ionospheric plasma swirl around the uppermost reaches of the atmosphere like thunderclouds and weather fronts. The flat ionosphere that was once useful for reflecting radio waves suddenly becomes rippled like a piece of corrugated cardboard. Light and radio waves get refracted (bent) in a phenomenon known as ionospheric scintillation (similar to the way light is refracted by water, such that a pencil looks bent when it is half-submerged in a glass of water). Predicting how, where, and when these distorted radio signals will bounce back to Earth becomes difficult. Since radio transmitters are calibrated to certain frequencies and conditions in the ionosphere—to certain bends and reflections of their signals—space weather scintillation can cause systems to lose their "lock" on certain frequencies.

At other times, the ionosphere can become so excited and dense that it absorbs the signals it usually reflects, causing faded signals and sometimes radio blackouts. During the March 1989 storm, many high-frequency radio channels were unusable for long periods of time. Shortwave radio fadeouts hampered commercial airlines, ship-to-shore radio, and international broadcasts by the BBC World Service, Radio Free Europe, and the Voice of America. In Minnesota, ham radio operators picked up the radio transmissions of the California Highway Patrol, some 1,500 miles away. Television viewers in Key West, Florida, reported watching one network and hearing the audio track from another. Some shortwave transmissions were interrupted for as long as 24 hours.

Similarly, the U.S. Coast Guard's Long-Range Navigation (LORAN) system was rendered nearly useless for several hours during the March 1989 storm when its very low-frequency signals were blocked due to the sharply increased density of the ionosphere. In a coastal town in California, automatic garage doors began opening and closing on their own. Some people wanted to blame the raging magnetic storm overhead, but it wasn't the Sun and aurora at work—at least not directly. A U.S. Navy ship cruising offshore had started using a special radio frequency to keep in communication with its coastal base. The only reason the vessel was using that frequency—nearly identical to the frequency used by garage door systems—was that the LORAN system was blacked out by the solar onslaught.

One of the ways to overcome the radio distortion in the ionosphere is to transmit signals at frequencies that pass through it, relaying radio messages through satellites instead of the less predictable atmosphere. And while that method is more effective and precise, it is exponentially more expensive and not without its own set of problems. On rare occasions, turbulence can make the ionosphere opaque to certain radio frequencies, preventing signals from passing through and cutting off communications between ground stations and spacecraft.

More often, space weather makes signals hard to track with precision. Scintillation can be a problem for some users of the Global Positioning System (GPS) and users of satellite phones and television networks. For instance, in 1997, operators working through a performance review for the Federal Aviation Administration lost their lock of GPS signals from four of their five stations during an excruciating 13 minutes. Systems such as guided missiles that depend on extremely precise tracking of radio signals can also have problems because scintillation changes the path of those signals. In a recent report the U.S. Department of Defense estimated that scintillation could cause as much as 27 minutes of delay in the command and control of its Tomahawk cruise missiles.

Even the Sun itself can get involved in breaking up radio communications, as scientists figured out during the tense days of World War II. In 1935, British physicist Robert Watson-Watt produced the first practical radar (radio detection and ranging), allowing operators to locate objects beyond their range of vision by bouncing radio waves against them. Radar can determine the presence and range of an object, its position in space, its size and shape, and its velocity and direction of motion. That was particularly useful for Britain as it struggled to fend off the bombers and fighter planes of the Nazi Luftwaffe.

Late in February 1942, British radar stations watching over the English Channel started picking up severe rushing noises. Their systems periodically became completely inoperative because of a very strong form of radio noise. Operators grew concerned that the Germans were jamming their system and a major attack was

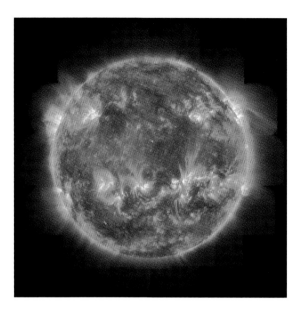

The Sun is an intensely magnetic body and that magnetism is revealed in the bright field lines and loops of plasma suspended in the atmosphere. NASA's TRACE satellite snapped the many images that make up this composite view of the Sun.
Courtesy of NASA.

Million-degree plasma hangs in looping magnetic field lines near the visible surface of the Sun. These loops stretch nearly 300,000 miles high. Scientists hope to learn how the solar atmosphere is heated to 300 times the temperature of the surface.
Courtesy of NASA.

This artist's conception shows the principal parts and layers of the Sun, using some real imagery from the SOHO satellite.
Image created by Steele Hill for ESA/NASA.

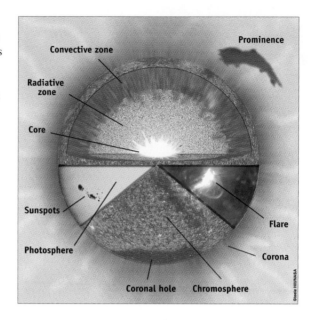

A large, eruptive prominence extends more than 35 Earth diameters above the surface of the Sun in July 1999. Viewed here in ultraviolet light by SOHO, prominences are dense clouds of plasma trapped in the Sun's magnetic field. Eruptions of these loops are often associated with solar flares and coronal mass ejections (CMEs). Courtesy of ESA and NASA.

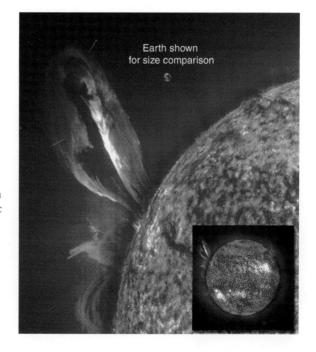

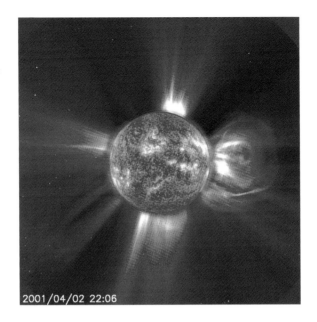

This image offers a composite view of the Sun in April 2002, when a CME lifted off (right side) and caused space storms on Earth. Since 1996, instruments on SOHO have enabled scientists to view the Sun 24 hours a day, creating new images every few minutes. Courtesy of ESA and NASA.

An artist's conception of a CME and its subsequent impact at Earth. The blue paths emanating from the Earth's poles represent some of its magnetic field lines, which deflect most of the energy and plasma of the CME. (Not drawn to scale.) Image created by Steele Hill for ESA/NASA.

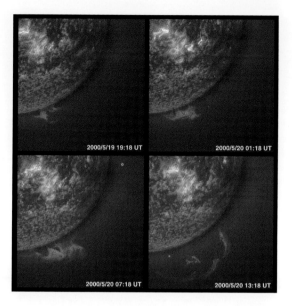

Over the course of 18 hours, a prominence grows from the limb of the Sun and erupts in a coronal mass ejection. A typical CME blasts billions of tons of particles into space at millions of miles per hour. Courtesy of ESA and NASA.

The time-lapse sequence of images from the Large-Angle Spectrometric Coronagraph (LASCO) on SOHO depicts a huge bubble of plasma as it is blown away from the Sun. In the last frame (bottom right), solar protons hit the camera, causing "snowy" interference. Courtesy of ESA and NASA.

Comet Hale-Bopp shows off its two tails in 1997—one of dust (white) and one of ions (blue). Comet tails always point away from the Sun, an observation that led to the discovey of the solar wind. Photo by Fred Espenak.

These images come from a computer simulation of the magnetosphere during the magnetic storm of January 1997. Earth lies within the sphere where the two axes meet. The scale in the upper right indicates the density of plasma. The shock wave—known as the bow shock—in front of the Earth is clearly visible, as is the boundary of the magnetosphere. On the lower right is a simulated auroral display in ultraviolet as viewed with the North Pole at the center. The region of magnetic field connected to Earth's North and South Poles is indicated by the comet-like surface. Note how the magnetosphere shrinks and auroras light up after the arrival of the solar storm. Courtesy of Ramon Lopez.

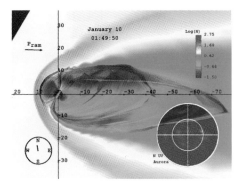

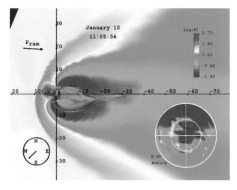

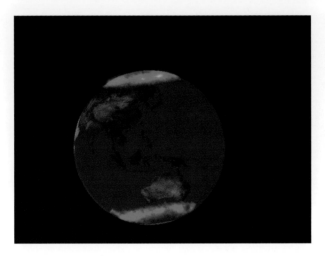

On October 22, 2001, the Visible Imaging System on NASA's Polar spacecraft captured both the *aurora borealis* and the *aurora australis* (northern and southern lights) in one image of Earth, only the third time such an image has ever been shot. For nearly three centuries, scientists have suspected that auroras in the northern and southern hemispheres were nearly mirror images of each other, but it was not until the Space Age that they were able to confirm that global perspective. Courtesy of VIS/ University of Iowa and NASA.

Astronauts on space shuttle flight STS-39 shot this photo of the *aurora australis,* or southern lights, in May 1991. Auroras form along the edge of space, where high-energy particles trapped in our planet's magnetic field slam into the gases of the upper atmosphere. Courtesy of NASA.

Visions such as this auroral serpent preparing to swallow the Moon make it easy to understand how ancient peoples could see monsters, bridges, and ghostly warriors in the northern lights. The photo was shot in November 1998 in Alaska. Courtesy of Dick Hutchinson.

Auroras can take several forms—curtains, rays, bands, arcs—and colors—mostly greens, reds, whites, purples, and blues. In this case, the aurora has formed a rare "corona," a phenomenon where the center of the aurora appears directly overhead, radiating from a center point. Courtesy of Jan Curtis.

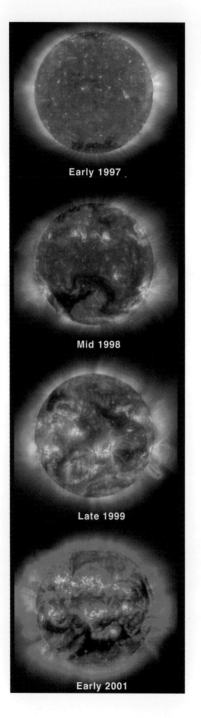

A comparison of four images of the Sun in ultraviolet light, taken by SOHO over the course of five years, illustrates how the level of solar activity has increased significantly with the arrival of solar maximum 23. Many more sunspots, solar flares, and coronal mass ejections occur during the solar maximum. Courtesy of European Space Agency and NASA.

imminent. A physicist in the British Army Operational Research Group, James Stanley Hey, was charged with investigating this jamming of Army radar sets. Hey and colleagues discovered that the Sun was a powerful and highly variable radio transmitter and that sunspots and other forms of solar activity were producing potent radio emissions. The enemy of the radar was not the Germans but the Sun.

The effect persists today. Many satellites are typically placed in geostationary orbits around Earth, orbits that lie in the equatorial plane. During the spring and fall equinoxes, the Sun also passes through this plane (from the perspective of Earth). Once a day the satellites pass in front of the Sun and their signals get lost in the radio noise.

8   A Tough Place to Work

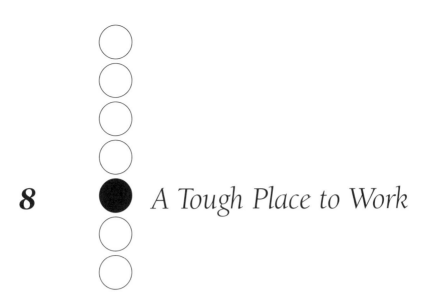

> Any sufficiently advanced technology is indistinguishable from magic.
>   Arthur C. Clarke's Third Law, *Profiles of the Future:*
>   *An Inquiry into the Limits of the Possible*

At least 600 satellites were whirling in orbit above the Earth at the start of the year 2001, an armada of commercial, military, and civilian science spacecraft totaling about $100 billion in public and private investment. Nearly 100 of those spacecraft relay military secrets and strategies, everything from images of enemy armies and missile silos to weather reports from remote battlegrounds of the world. Another subset of those 600 satellites gathers scientific data for government agencies, watching cloud patterns, whale migration, ozone holes, and the exploding stars in far-off galaxies.

The rest belong to a burgeoning industry—satellite communications. Private firms having been using satellites for the delivery of mass media for many years, with radio and television networks—even those "cable" networks—beaming their programs via satellite to the local dishes of their affiliates. Telephone and paging services are also traditional customers of the satellite industry. But today, satellites also relay stock indexes, transactions from automated teller machines, retail inventory counts, truck movements, and consumer payment information at some pay-at-the-pump gasoline stations. Satellites have become a major yet invisible part of the commerce of modern economies. Most of us don't realize how much our lives are intertwined with satellites until our pagers go silent or our favorite television programs are blacked out. And most of us know even less about the intricacies of using satellites than we do about how oil, gas, wind, or coal gets turned into the electricity coming into our house.

It took more than three decades for humans to put those 600 active satellites into orbit, not to mention the thousands of spacecraft that have already expired. According to the Teal Group, an aerospace consulting firm, more than 2,100 new spacecraft have been proposed for launch by the year 2010, at a cost of at least $220 billion. About 1,000 of those payloads will be commercial communications satellites.

Yet just as all of these satellites are queuing for takeoff, the failure rate for launch vehicles and satellites has increased dramatically, according to Christopher Kunstadter, senior vice president of U.S. Aviation Underwriters, a major satellite insurance company. The long-term failure rate for all launches is roughly 10 percent, he notes, but in the first 10 launches of a new type of rocket, the rate jumps to 20 percent. And the problems don't stop at the launch pad and rocket booster. Space weather has been blamed as the cause or contributor to more than $500 million in insurance claims between 1997 and 2001, and in numerous other cases, satellite companies lost some of their capacity and redundancy in mishaps that could not be claimed. Since January 1998, the insurers of satellites have paid more than $3.8 billion for 22 total mission losses and 91 partial losses; uninsured losses over the same period exceeded that amount. In the grimmest of recent years, 1998 and

2000, the value of insurance claims exceeded the amount of money collected in premiums.

"The space insurance market is reeling from recent losses," notes Kunstadter. "These losses reflect a real decrease in space system reliability. As a result, available capacity [for insurance] has decreased, rates have increased for all types of coverages and risks, and general terms and conditions have tightened."

Furthermore, commercial pressures in this fiercely competitive market are forcing companies to fly satellites that may not be as hardy as the satellites of the past. "Cost and schedule drive technical considerations," Kunstadter says, so things like spacecraft testing are being reduced in order to meet schedules. The high cost of radiation shielding for satellite parts also means that—due mainly to financial pressures—most satellites are less sheltered today from space weather than they were in the past.

Most satellite failures occur early, either on the launch pad or shortly after leaving it; others fail due to the breakdown of key parts once the spacecraft is out of reach in space. Orbital debris and meteors take out a few satellites. But experts cannot agree on just how many perish due to space weather. The Aerospace Corporation, a leading space technology research firm in the United States, claims that about 5 percent of all spacecraft failures can be attributed to the environment; the cause of another 10 percent of mishaps is "unknown." According to Alan Tribble, who has been teaching courses on space weather effects since 1992, that figure for environmental failures might be as high as 25 percent, if you account for the long-term degradation of solar cells and other spacecraft parts. Dave Speich, a recently retired forecaster for the National Oceanic and Atmospheric Administration's (NOAA) Space Environment Center (SEC) estimates that one-third of all spacecraft anomalies—including failures and corrected problems—can be attributed to the brutal environment of space.

The trouble with assessing the impact of space weather is that without collecting the satellite and looking under the hood to see what went wrong—something that is logistically and financially impossible for all but a few NASA satellites—it is difficult to know with certainty just what caused a given spacecraft to fail. Satellites deliver reams of data on the health of their electronic systems and the local space environment, and that information is scrutinized down to the millisecond. Satellite companies regularly reconstruct in fine detail the exact sequence of events before a spacecraft anomaly or failure. But such data and the analysis of it are closely guarded secrets in the satellite industry, proprietary matters that are rarely shared with scientists, much less competitors.

So whenever a satellite dies in the middle of a space weather event, many space scientists—and a fair number of engineers from competing companies—speculate about whether the environment had something to do with it. Maybe a storm wiped out a satellite or perhaps it was the trigger event that exposed a more fundamental problem with the design of the spacecraft (such as the "tin whiskers" of Galaxy IV). The predicament is that these outsiders—some of them with an agenda, others with a lifelong scientific interest—can only cite circumstantial evidence, like trying a murder case without a witness. Unless they work for the company that suffered the loss, they don't have access to the data to prove how space weather may have killed any satellite. On the other hand, those same speculators are rarely refuted or challenged in public. It is a puzzling and intriguing dance between scientists who understand the space environment and engineers who know their satellites, between researchers bucking for the next publicly funded grant and researchers protecting their company's future.

On January 13, 1994, the Sun began spraying high-speed solar wind at Earth. A hole in the corona was allowing the fastest breed of solar wind—usually two to three times faster than the everyday variety—to escape the Sun's intense magnetic fields and rush across the solar system. At Earth a minor magnetic storm began and proceeded rather humbly until January 19. NOAA's Space Environment Center noted the storm but did not feel compelled to issue any alerts. It was, by most measures, a relatively modest event.

Modest or not, it was enough to disable a satellite. At about 12:40 p.m. on January 20, the Anik E1 satellite began to spin out of control. Operated by Telesat Canada, the three-year-old, $300 million satellite lost contact with ground stations when a failure occurred in the primary attitude control circuit that kept the spacecraft oriented and pointed in its orbit over the Americas. Fortunately, controllers were able to resolve the problem within about eight hours by switching to the backup system. E1 was pointed back at Earth and ready for business.

Its twin satellite was not so fortunate. Within an hour after communication with Anik E1 was restored, Anik E2 went into a similar tumble. The attitude control system had a failure much like the one that had temporarily felled E1. But when controllers switched to the backup system for E2, they found that it was dead, too. Anik E2 was useless and engineers struggled for months to restore the satellite.

The failure of two of Canada's prime satellites caused communications havoc across the country. Several hundred of Telesat's corporate customers were affected by the outage, including stock quote services, retail sales inventory systems, and scientific data relays. Long-distance telephone service was cut off in the northern reaches of Quebec and Ontario and all of the Yukon and Northwest territories. But the hardest-hit sector was the media. More than 100 newspapers, radio stations, and television networks were sent scrambling for alternative links on January 20. Many newspapers were using satellites to distribute stories and photos to their affiliates and to relay their pages to printing presses around the country; suddenly they were using faxes and film. The Canadian Broadcast Company could not distribute its Newsworld programming to 450 affiliates, and most of its specialty cable programming went black for the day.

Within 24 hours of the failures, most of Telesat's customers had some form of satellite service restored. Many of the customers of the E2 satellite were switched to E1, while others were temporarily moved to Hughes' Galaxy VI satellite. Eventually Telesat leased space on AT&T's Telstar 301 to satisfy all of the remaining

E2 customers. Seven months later Anik E2 was resurrected to useful life when engineers developed a complicated system for controlling the satellite's thrusters from the ground. After all the damage was done, Telesat forfeited between $50 million and $70 million in lost revenues and recovery expenses. The Anik E2 satellite lost a year and a half of useful life: six months spinning out of control and one year taken out of its life expectancy due to the amount of fuel that now has to be used to keep the satellite pointed in the right direction. The operating cost for the rest of E2's life grew by $30 million.

Spacecraft operators and the forecasters at SEC were initially puzzled by the Anik failures. Telesat claimed that the environment was "100 times worse" than anything it had seen before, but a first glance at the usual scientific data did not support such a claim. As magnetic storms and solar wind streams go, the conditions were more active than normal, but nothing at all like the major storm conditions of 1989. The intensity of the energy coursing through Earth's magnetosphere did not seem to implicate space weather as the culprit.

"Within a day of the Anik failures, I was called by representatives of three different groups," says Joe Allen, former head of the solar-terrestrial physics division of NOAA's National Geophysical Data Center. "One was someone from Telesat Canada; another was an official in the Canadian Ministry of Defense; the third was a vice president of the bank that had financed the building of the two satellites. All wanted to know whether the cause of failure could have been 'sabotage,' coincidence, or a result of a disturbed space environment. The U.S. Air Force facility at Falcon Air Force Base in Colorado had already run an analysis of the event and concluded with 'high confidence' that there was a 'low probability' that the failures were not due to natural causes."

"I brought down the data on my computer and looked at the energetic electrons for the previous week," Allen recalls. "The daily variation of the high-energy electrons looked like a textbook illustration of a smooth sine wave, they were so regular. I called Dave Speich to talk about it and told him I didn't believe such satellite anomalies could happen without energetic electrons being involved. Dave looked at the data and then said 'Oh, my goodness.' We both noticed then that while the daily variation looked like what you would usually expect, it was two to three times higher than normal. This condition had persisted for almost 10 days. It was not a 'storm' per se, although a Canadian university group mistakenly announced that it was." But there was little denying that the environment had somehow damaged the satellites.

In addition to the Anik troubles, the Intelsat-K international communications satellite wobbled a bit on the twentieth, and the GOES (Geostationary Operational Environmental Satellites) and several spacecraft endured minor problems that day. "There was just so much compelling evidence that the environment had to be involved," says Dan Baker, a University of Colorado space physicist who has studied the Anik problems. Closer analysis in the months after the failures revealed that the amount of high-energy electrons drifting in the radiation belts (around the time of the failures) had increased by a thousand fold and stayed that way for nearly eight days. The high-speed streams of solar wind buffeted the magnetosphere to a point where it began accelerating the particles trapped in the Earth's space environment. Though it was not necessarily the sort of change that could kill a satellite instantly, Baker notes, the sustained increase in "killer electrons" probably inflicted a creeping death. It is known as deep dielectric or "bulk" charging.

Dielectric materials are supposed to be a spacecraft's best friend, and usually they are. But sometimes they foster the most shocking of space weather effects. Dielectric materials are insulators—that is, materials that prevent the flow of electricity. Dielectric materials are used to insulate circuits and cables from each other deep inside the machinery, making sure currents flow where they are supposed to and do not disturb other components. Even when some of the highest-energy electrons from space penetrate a spacecraft and lodge themselves in the dielectric insulation, those electrons leak away over time at a steady rate that keeps the satellite safe.

But when a spacecraft is immersed in a long-lasting and intense bath of high-energy particles, it can be overwhelmed. Because dielectric materials are designed to keep out most electric charges, the particles that do manage to get embedded in the insulation are inevitably particles of the highest energy. The accumulation of excessive electric charge in dielectric insulation is extremely rare, but when it does occur, it is often catastrophic. The effect is comparable to water flowing into a bathtub. During normal flows, the drain keeps the water level low. But if the flow suddenly increases, the water level rises until a new equilibrium is reached between the flow of water (electrons) in and the flow out (leakage from dielectric materials or metal components to an electrical ground). If the flow gets very high and stays high for long enough, the tub overflows. If the electron flux is high enough and persists for long enough, then electrical discharges occur that cause satellites to misbehave and sometimes fail. Scientists who studied the space environment during the failures of the Aniks suspect that deep dielectric charging was the culprit. The environment around the satellites was so intensely charged for so long that the buildup of electrons in the insulation provoked debilitating sparks and shocks.

Whereas deep dielectric charging by electrons is the most insidious and perhaps most destructive space weather effect, the most common form of disturbance is the single-event upset (SEU). SEUs are typically caused by high-energy ions, which are 1,800 times more massive than electrons. Accelerated to damaging speeds in the radiation belts around Earth, these particles can pierce a spacecraft and its components, literally punching holes through them. SEUs usually involve changes to the memory of a satellite's computer circuits, sometimes locking up the electronic brain temporarily in the same way a terrestrial computer can crash. Other times the high-energy particles cause satellite sensors to give false readings (see Figure 13). Statistics from NOAA-SEC show that more than 950 SEUs were reported during the last solar cycle (roughly 1985 to 1995), about one for every spacecraft in orbit per

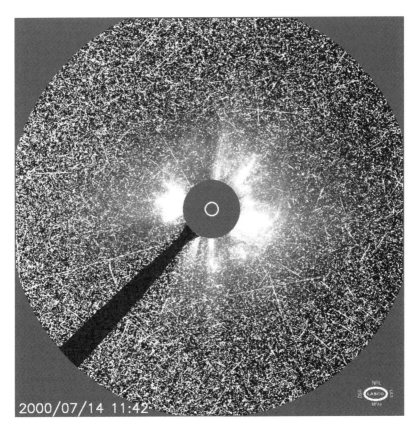

FIGURE 13. An image from the Large-Angle Spectrometric Coronagraph (LASCO) on the SOHO satellite shows solar protons bombarding and overwhelming the spacecraft's instruments. SOHO and other satellites occasionally lose the use of their cameras when swarms of solar particles "white out" the view like interference "snow" on a television. Courtesy of SOHO/ European Space Agency and NASA.

year (though a few satellites actually endured the majority of those upsets and many more SEUs likely went unreported).

The Hubble Space Telescope is a famous victim of single-event upsets, though many satellites (including the Global Positioning System for navigation) have experienced them. Hubble's orbit regularly takes it through the "South Atlantic Anomaly" (SAA), which is located over South America and the Atlantic Ocean. The SAA is a region where the Earth's magnetic field takes a sudden "dip" in intensity. This allows high-energy particles from the Van Allen radiation belts to reach lower than normal altitudes, causing problems for any spacecraft that flies through the SAA. One instrument on Hubble cannot operate when the satellite is passing through the region because an SEU could reset the instrument to its highest voltage and perhaps shock the instrument to death. That detector must be shut off for a few minutes every time Hubble heads for the South Atlantic, about 7 out of 16 orbits per day.

In 1999, two brand new NASA satellites—the Far Ultraviolet Spectroscopic Explorer (FUSE) and the Terra earth-observing spacecraft—suffered from corrupted memory and single-event upsets due to high-energy radiation from the radiation belts, particularly the South Atlantic Anomaly. As with Hubble, detectors on the new spacecraft sometimes have to be shut down when passing through that high-energy region. "It's a ferocious radiation environment up there," said Johns Hopkins physicist and FUSE principal investigator Warren Moos in a newspaper interview. "But I think the choice of [computer] chip was not the best." Tougher, "radiation-hardened" chips could shield the spacecraft's memory for decades. But as with the rest of the satellite industry, NASA is on a tight budget and a tighter schedule than in years past. Deputy project scientist Kenneth Sembach told Space.com that the FUSE engineers "chose the best they could given the costs and the time constraints we were under."

Another space weather problem is known as surface charging, which can lead to a damaging "electrostatic discharge" on the outside of the spacecraft. As a satellite drifts through the plasma trapped in the space around Earth, it tends to accumulate electrons on its exposed surfaces. It is much the same as the way you pick up extra charge and shock yourself when you drag your feet across a carpet in the wintertime. Many spacecraft are designed with conducting surfaces to bleed off that accumulation of charges or direct it into grounding devices. But sometimes the flow of electrons is too much for the safeguards. The satellite lights up with an electrostatic shock—space lightning—that can burn out a power supply or shock the components connected to the surface of the spacecraft. Sometimes these sparks can look like an electromagnetic signal from the ground and trick the satellite into following a phantom command.

In 1991 a potent electrical shock to the solar panels crippled the 10-year-old Marecs-A navigation satellite as it hovered over the Atlantic. In the case of another spacecraft, one of NOAA's Polar Orbiting Environmental Satellites, the official cause of death was listed as a bolt sticking out of the solar panels. More than just a symbol of poor design and construction, the bolt became a sort of attractor for electrons. The bolts holding the solar panels together were supposed to have been insulated with a layer of Teflon. But the bolts were too long, so they pierced the Teflon coating during construction. After launch, the satellite flew without incident for two weeks until the radiation belts grew active. As electrons accumulated on the bolts, a lightning-like shock short-circuited NOAA-13's solar power systems.

The energetic particles floating in the space around Earth also can damage spacecraft without causing immediate catastrophic failures. Instead, the destruction is slow, cumulative, and permanent. The steady dose of radioactive particles can destroy the principal energy source for most spacecraft: solar panels. Designed to absorb sunlight and turn it into cheap power for spacecraft, solar cells also collect a lot of other radiation from flares, coronal mass ejections (CMEs), galactic cosmic rays, or just the steady streaming of particles trapped around the Earth. High-speed particles smash into the semiconductor materials and reduce the expected lifetime over which the solar cells will be able to produce energy.

In October 1989, following a particularly intense flare that sent a swarm of high-speed protons toward Earth, at least 13 geosynchronous satellites suffered permanent damage to their solar cells, and one of the GOES weather satellites lost six years off of its designed life. A comparable storm in 1991 took two years of life from each of NOAA's three GOES weather satellites by reducing their power-generating capacity. Over the course of its 15 years in orbit, the *Mir* space station suffered power shortages due to the decay of its solar panels. And as recently as 1997, two newly designed satellites lost nearly one-fifth of their power-generating capacity after just two solar storms.

Space weather also has an indirect effect on satellites through increased atmospheric drag, or friction. Intense space weather events can heat the upper atmosphere of Earth, inflating it so that the gases around satellites in low-Earth orbit (LEO: altitude 300 to 500 kilometers, 180 to 300 miles) become denser. This expansion of the atmosphere significantly increases the number of microscopic collisions between the satellite and the gases and plasma of the upper atmosphere. The increased friction, known as "satellite drag," can alter an orbit enough that the satellite is temporarily "lost" to communications links. It can also cause the premature decay of the orbit; that is, it can cause a satellite to fall out of orbit and burn up in the atmosphere sooner than planned. The Hubble Space Telescope has been boosted back up to a higher orbit during the servicing missions in order to counteract atmospheric drag. On the other hand, atmospheric drag from solar maximum 21 caused NASA's Skylab space station to crash back down to Earth in 1979 before the first space shuttle could be launched to boost it into a more stable orbit.

The U.S. Department of Defense (DoD) spends roughly $500 million per year to mitigate the effects of the space environment. Those effects range from outright failures, to the loss of service time due to satellite degradation, to time spent correcting computer failures or satellite positioning. From 1987 to 1991—the height of the last solar maximum—DoD had to respond to as many as 300 satellite "anomalies" per year. Even in the years of solar minimum from 1994 to 1996, the military still dealt with 150 satellite problems per year.

With more than 30 communications and missile-warning satellites operating in geosynchronous orbit, 25 GPS (Global

Positioning System) spacecraft flying through the radiation belts, and 25 weather and science satellites flying in low-Earth orbit—not to mention the many classified reconnaissance spacecraft—the U.S. Department of Defense has strong reason to be concerned about space weather. In preparation for the current solar maximum and anticipating future space-based defense systems, the U.S. government's National Security Space Architecture office gathered scientists, engineers, and decision makers from all branches of the military, civilian science agencies, and industry to conduct a national Space Weather Architecture Study. The group surveyed existing science and satellite data and projected that satellite operations could be disrupted as much as 15 percent of the time during solar maximum (2000-2002), with radio transmissions intermittently interrupted about 20 percent of the time. They also predicted GPS navigation errors about 20 percent of the time.

The vulnerability of the military is not necessarily an issue of hardware. Though some military satellites are using newer, less radiation-hardened parts, most of the crucial DoD satellites have been built to withstand the radiation of a nuclear weapon exploded in space. "Space weather is a threat, but not a day-to-day threat," notes Chris Tschan, a retired Air Force officer who commanded the Air Force's 55th Space Weather Squadron for a few years in the 1990s. "It is more of a nuisance to the military."

The real problem is ignorance. Many of the officers and enlisted soldiers operating DoD spacecraft, communications devices, and navigation systems have scarcely heard of scintillation, CMEs, or solar wind. The Space Weather Architecture Study revealed that military users frequently don't understand space weather effects and their operational impacts are not well documented for the average soldier. Out of those two problems spring a host of others, such as the fact that the current military requirements for space weather information and forecasts are outdated, fragmented, and incomplete. Military and civilian space weather requirements are similar but often addressed independently. And the lack of fundamental knowledge about the physics of space weather and its impact on technology makes it difficult to develop and evaluate techniques to compensate for space weather.

"In the 1990s I observed that space system operators did not understand the magnitude or manifestations of the space environment on the systems they operated," Tschan recalls from his days in the Air Force. "Most had a one-page summary, circa the 1960s, that told them there could be impacts. Since then I think the level of awareness of space environmental impacts is up substantially, but I feel confident that operators are still weak in knowledge about how to handle a space environment-caused anomaly."

Tschan remembers an occasion when there was a partial radio blackout and false missile launch indications on the warning radar systems. The signals were caused by aurora, but the radar operators had no idea that the northern lights could cause such a signal. So they essentially ignored radar signal returns from that portion of the radar display, even though Air Force One was in flight at the time.

"An Air Force officer once remarked to me, 'I can't remember losing any spacecraft to space weather,'" Tschan recalls. That's because many of the long-term effects of the space environment are hidden by the fact that DoD replaces many spacecraft before they have a chance to fail. "Basically, the space operators compensate for failing systems and switch to redundant systems, or they nurse ailing space systems along more often than you'd care to know, masking the problem to some extent."

Lack of information about space weather is making the military vulnerable in other ways, too. As the federal budget shrinks and the military tries to get more for its dollars, the Pentagon is leasing time on commercial spacecraft and buying more satellites with off-the-shelf technology—that is, satellites that commercial firms already produce, instead of custom-made spacecraft and components. At the same time, only two companies are producing radiation-hardened parts, down from 20 suppliers in the Cold War era.[1] Designers are installing less costly commercial computer processors in missiles and satellites because they weigh and cost much less than the bulky, radiation-hardened chips that DoD has used in the past. Many officers who command or rely on space technology probably do not even realize that their budget savings may have a cost later. Others may be aware but do not have the money to do anything about it.

"I believe that military leaders are more aware of space environmental effects than they have ever been," said Tschan. "But my feeling is that it isn't enough, especially for systems with smaller-scale electronics that haven't weathered a solar maximum before. Space weather is still something they don't really understand because it isn't tangible, like a thunderstorm or tornado. I'm sorry to say, it may take a small catastrophe for the vulnerability issue to sink in."

On January 6, 1997, the Sun blasted a coronal mass ejection in the direction of Earth. By January 10 the cloud from the Sun was compressing Earth's magnetosphere, stirring up energetic particles in the radiation belts and ionosphere and causing a major magnetic storm. By January 11 a telecommunications satellite died. Were the events connected?

AT&T announced to its customers and the news media on January 11 that its $200 million Telstar 401 satellite had "experienced an abrupt failure of its telemetry and communications" at 6 a.m. that morning. The satellite's main control systems suffered a catastrophic failure, and the satellite went "out of orbital alignment," the official statements noted. Four of the major television broadcast networks in the United States—ABC, Fox, UPN, and PBS—lost the ability to transmit programming to their affiliates. For Fox the failure couldn't have come at a worse time, as the network was preparing to broadcast the NFC football championship the next day. Every PBS station in the country had received its programming via Telstar 401, so every station had to readjust its satellite dishes as AT&T found new relays for their signals. The satellite also was the primary relay for close to half of the syndicated television shows at the time, including "Oprah,"

"The Simpsons," "Wheel of Fortune," "Baywatch," and several others.

Part of the U.S. earthquake-monitoring network also was disrupted by the loss of Telstar 401. The U.S. Geological Survey (USGS) had used the satellite to transmit near real-time data about seismic activity to the agency's field centers. Though the regional earthquake networks in the more active regions—California, the Pacific Northwest, Alaska, and Hawaii—were not affected, almost everything east of Nevada was knocked out of service. USGS acknowledged that its reporting time for earthquake data for two-thirds of the country was slowed by several hours.

The failure of the satellite ultimately turned out to be more of an inconvenience than a catastrophe for the media outlets. While operators worked to restore contact with the silent satellite, AT&T quickly transferred service for most customers to another satellite, Telstar 402R, as was spelled out in contractual contingency plans. Some customers, such as USGS, had to wait a bit longer, as their contracts were considered lower-priority items. Within six days, AT&T declared that Telstar 401 was "permanently out of service," and AT&T Skynet began making plans to move another satellite, Telstar 302, into the defunct satellite's slot in space.

The loss of Telstar 401 was a bit more than just an inconvenience for AT&T. In the fall of 1996, AT&T had agreed to sell its Skynet Satellite Services to Loral Space & Communications for a total of $712 million. With the loss of Telstar 401—which had been expected to last for another nine years—AT&T was forced to renegotiate. When Skynet finally changed corporate hands in March 1997, Loral paid $478 million. AT&T also collected $132 million in insurance claims for the loss of the satellite.

Lost in the shuffle of inconvenienced customers and humbled executives was the true fate of the satellite itself. AT&T announced at the time that a team of spacecraft experts from AT&T and Lockheed Martin, the manufacturer of the satellite, was being assembled "to determine the root cause of the problem." But with Loral planning to build its own satellites and AT&T getting out of the business, neither company had a strong interest or obligation to report on the results of the investigation. Louis Lanzerotti, a physicist for AT&T's research cousin, Lucent Technologies, told *Science News* at the time that the satellite could have failed for several reasons, but "the coincidence with the magnetic storm is uncanny." Since that time, nearly every scientist in the space weather community assumes and asserts that Telstar died from the one-two solar punch of January 1997. But none of them really know for sure.

There is an extremely low rate of failure due to space weather, at least by official counts. "Hundreds of satellites have functioned normally for close to a decade, providing an incredible financial rate of return for their operators," notes one industry insider who prefers to remain anonymous. "The only reason that spacecraft failures make the news is that they are rare to the point of being shocking when they occur. The interruption in cable TV service due to power outages from terrestrial storms—or even something as mundane as a backhoe operator accidentally cutting a cable—never makes the news because we are familiar with and even expect bad weather and its inconveniences."

But some scientists wonder if the space weather problems are rare or just rarely reported. Satellite operators and builders are under tremendous economic pressure not to report problems. Satellite builders certainly don't want to scare away future customers or investors by admitting they might be vulnerable to space weather. Anomaly reports are privileged private information for the companies that build and own the satellites. Most of the parties involved—even the insurers—are bound by confidentiality agreements. "If the news gets out that a satellite might have failed due to space weather, that makes customers extremely nervous," says one industry source. The cause of a failure is not necessarily information that companies care to share with investors and customers—and competitors—in a fiercely contested market.

"They don't want to admit that there is a problem because it would create a competitive disadvantage," says Joe Allen, who has long been the scientific community's unofficial collector of anecdotes and data about spacecraft troubles. "Industry knows a great deal about the conditions in space and what happens. They say, 'We know there are problems. We just don't talk about it.' It would affect the cost of their vehicles."[2]

Some failures have been reported, principally the failures of government satellites. According to the Space Weather Architecture Study, 15 satellites have failed as a direct result of space weather since 1983, and those failures are spread out across the entire solar cycle, from active solar maxima to relatively quiet minima. Eight of those losses befell the first satellite in a new series, underscoring the unknowns in the development of new spacecraft. Chronic degradation by space weather also led to substantial redesigns of at least 12 major satellite systems in the past 20 years, and at least 21 satellites have had power supply problems that were caused by the environment. The authors of the architecture study summed it up for the skeptics in three words: "Silver bullets happen."

Some industry watchers believe there could be a lot more silver bullets in the next few years. The race to claim a share of the space market has led many companies to build satellites with the newest untested technology or with less radiation hardening, which can double the cost of a satellite. "Budget and competitive pressures, as well as increases in demand for improved coverage, timeliness, accuracy, and assuredness in space-based service, are likely to make future space weather impacts more significant," the authors of the architecture study concluded. Add in that the number of reported satellite anomalies traditionally doubles in years of major magnetic activity, according to Allen, and the current solar cycle could get interesting.

"A catastrophe is more likely for civilian systems since there are more of them and most have newer designs," Chris Tschan notes. Most spacecraft manufacturers "overengineer" their systems because they don't know exactly how much radiation their satellite will face. In addition, spacecraft components tend to be tested individually for radiation hardness. But those parts are not necessarily tested in the context of an integrated, complete system. Then there is the problem of protecting a spacecraft from radiation. It costs about $40,000 for every pound that gets launched into orbit. Any radiation shielding beyond the minimum required to protect parts for the forecast lifetime is considered dead weight. Spacecraft designers are conscious of the need to squeeze the fat out of their designs, so radiation hardening seems to be a luxury most satellite builders cannot afford.

Building a hardy spacecraft is like building a bridge or a house to withstand an earthquake, says one industry engineer. "The daily space weather is not so critical. It's the extremes that are important." But as in the military, many of the people who design or operate commercial spacecraft today have never seen the extremes and scarcely know what to look for. The models of Earth's radiation belts used to calculate environmental radiation doses are 20 to 30 years old. And the designers have so few case studies of failures or "anomalies" to study, unless their company has been one of the unfortunate victims of a storm from the Sun.

"I think space weather is a credible problem but not the only one to worry about," notes Alan Tribble, author of one of the only textbooks on space weather effects. "I don't feel that there are many glaring vulnerabilities in the average satellite, whether commercial or military. And as a rule, I think space weather does get appropriate attention in the industry. But most companies are prepared for the 'usual' environment. It's the once-a-decade kind of space weather phenomena that always give problems, and it's the same way on the ground. Hurricane Floyd [September 1999] caused problems on the East Coast because it was a very rare phenomenon. Things like that happen only a few days every few years. You either take your chances, or you drive up the cost dramatically by trying to make things hardened to survive the possibility of a worst-case environment."

"It is generally more cost effective to offset satellite problems with redundancy than to try for absolute perfection," notes one industry insider. "Satellite operators will purposely carry extra capability on the fleet of satellites they operate, so if failures occur they can shift traffic to backup capacity on other satellites. We all have spare tires in our cars since no one knows how, at a reasonable cost, to build a flat-proof tire that lasts 40,000 to 80,000 miles."

"It's very difficult to quantify how susceptible things are to space weather failure," Tribble adds. "I believe it boils down to a business decision. Solving the space weather problem has a cost associated with it. The return needs to exceed the investment. It's a matter of risk."

The simple fact is that no matter how expensive the hardware, no matter how commonplace the use of satellites, it will always be difficult and expensive to work in space. So is it negligent or just good business sense to take some risks with space weather? Satellite builders and operators are likely to find out through experience in the next decade.

9 Houston, We Could Have a Problem

Though thou art far away, thy rays are on earth;
Though thou art in their faces, no one knows thy going.
　　　Egyptian Pharaoh Akhenaten, "The Great Hymn to the Aten"

It was supposed to be solar minimum, the time when the Sun rests. But Nature does not heed human schedules; it makes them. In August 1972 the Sun produced a "sudden and spectacular resurgence of solar activity," as the staff at the National Oceanic and Atmospheric Administration's (NOAA) Space Environment Center called it. It was just a matter of luck that Apollo astronauts were not caught up in that resurgence.

On July 29, 1972, an average-sized sunspot group spun into view over the eastern edge of the Sun. While not unusually large, the region was magnetically complex, with snarled and twisted fields and intense gradients. That tightly wound knot of energy began to come undone on August 2, when the region exploded with three separate flares, two of them of the most potent "X-class." By the latter part of August 2, radio noise from the Sun reached a crescendo, making radio communications on Earth much more difficult, and next to impossible at Arctic and Antarctic stations. This pattern would continue for much of the next week.

August 2 marked the beginning of 10 days of flares and explosions that would stir up the magnetosphere and radiation belts of Earth. The series of flares produced a steady shower of "solar protons," hydrogen nuclei that are accelerated to extremely high energies by the solar blast. During solar proton events, these energetic particles travel from the Sun to the Earth within 20 to 30 minutes, and they can stream into the magnetosphere for hours. Upon reaching Earth, some of these particles spiral down Earth's magnetic field lines, reaching the upper layers of the atmosphere.

Close behind the proton blast, a coronal mass ejection (CME) raced toward Earth, though no one knew it at the time because CMEs had not yet been discovered. A magnetic storm commenced around Earth on August 4 and auroras dove into the continental United States in the early morning hours, with official reports from Colorado, Oregon, Vermont, Pennsylvania, and Illinois. By early evening, AT&T reported that one of its underground long-distance phone cables between Chicago and Nebraska was knocked out of service. Power companies in Minnesota, Wisconsin, South Dakota, and Newfoundland suffered through tripped transformers and fluctuating currents in their lines. And to top it off, the Sun released another X-class flare late in the day to further fuel the turbulent space weather. On August 5, communication through several of the transatlantic cables was interrupted.

The largest and greatest flare of the entire solar cycle lit up the Sun on August 7. The mangled sunspot region generated a flare that lasted at least four hours (see Figure 14). Radio signals from the Sun screamed at hundreds to thousands of times higher than normal levels, and the X rays and energetic particles saturated the sensors on several spacecraft such that the peak of the event could not be measured. Protons from the solar blast rushed to Earth and bombarded the upper atmosphere. In the Canadian province of

Houston, We Could Have a Problem

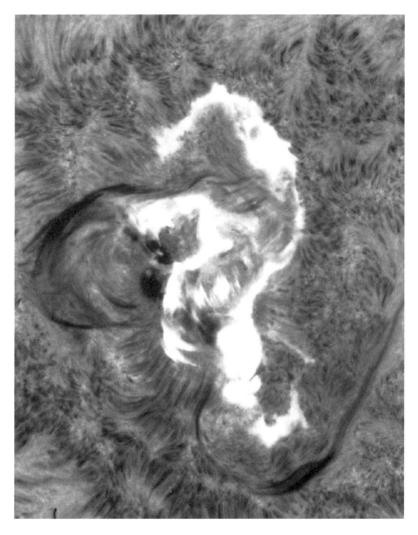

FIGURE 14. The great solar flare of August 7, 1972, became known to solar physicists as the "seahorse flare" due to its unusual shape. The intense radiation and swarm of particles from the flare probably would have been fatal to astronauts of the Apollo missions had they been in space at the time. Courtesy of Big Bear Solar Observatory/New Jersey Institute of Technology.

British Columbia, a 230,000-volt transformer blew up, the result of too many days of too much magnetic storming around Earth.

By the time the last major flare shot out from the Sun on August 11, Earth had endured the most intense swarm of solar protons since the era of satellite measurements began. At least 12 power companies had suffered mishaps or damage from the storm. Brilliant auroras reached as far south as Washington, D.C. But perhaps the most important effect of the storm was not one reported by power companies, satellite operators, or auroral observers. It was noted—and keenly studied—inside NASA.

By pure good fortune, the August 1972 flares and the string of solar proton events fell right between NASA's Apollo 16 (April 16 to 27) and Apollo 17 (December 7 to 19) missions to the Moon. With a simple change of launch dates, which can happen often in manned space launches, the astronauts could have easily been caught in the middle of the proton swarm. And for nearly three decades since the August 1972 event, scientists and flight surgeons have analyzed and reanalyzed the physics and the biology of the event, replaying it hypothetically in their minds and their computers. What if the astronauts had been outside of Earth's protective magnetosphere and on their way to the Moon when the Sun acted up? How much radiation would the astronauts have absorbed during the event, and how would it have changed the mission? What most of those scientists have concluded is that the astronauts might not have survived the trip.

"Although a great deal of thought and effort had gone into planning the Apollo missions, it was mostly luck that the crew was not subjected to excessive radiation exposure," said Gautam Badhwar, NASA's former chief research scientist for space radiation at the Johnson Space Center.[1] "If they were in the command module, it would have been dangerous but not life threatening. But there was not much shielding in the lunar module, probably not much better than a space suit. If the astronauts had been on the Moon, they probably would have been instructed to find a crater and dig down into it, as there is no magnetic field to protect them on the Moon."

In one study commissioned by NASA, researchers modeled the effects of the proton-accelerating flare of August 4, which began at 6:20 Universal Time (UT, more commonly known as Greenwich Mean Time). Within seven hours the Apollo astronauts would have absorbed the 30-day limit of radiation to their skin and eyes. Within eight hours the solar protons would have doused the crew with its 30-day limit of radiation for "blood-forming" internal organs and would have exceeded the yearly limit for exposure to the eyes. The limit for skin exposure would have been surpassed within nine hours, and one hour later the limit for a radiation dose to their organs would have been eclipsed. Within 11 hours after the flare the hypothetical Apollo astronauts would have exceeded the acceptable level of irradiation of their skin for an entire career.

In another study using measurements from the Solar Proton Monitoring Experiment on NASA's Explorer 41 satellite, researchers found that the August 4 flare raised the radiation dosage for an astronaut outside the spacecraft at one point to 241 rem per hour, and it sustained a level of at least 45 rem per hour for more than half a day.² Inside the Apollo command module, the rate would have approached 66 rem per hour. With a typical lunar round-trip lasting 11 to 14 days, each astronaut would have received a projected cumulative skin dose of 358 rem if he had stayed within the command module. According to the best projections available in 1972, astronauts who absorbed an acute dose of 340 to 420 rem over such a short period of time likely would have suffered from vomiting, nausea, and other symptoms of radiation sickness. At worst, each crew member would have been hospitalized for three to six months, with a 20 percent chance that they would have died from the accumulated radiation.

Radiation is one of the primary hazards of work in space. Regardless of storms from the Sun, astronauts receive a certain dose of radiation every time they go up, as the Earth is constantly bombarded by galactic cosmic rays from explosions and other phenomena that occur well beyond our solar system. They also receive small doses from the radiation that is naturally trapped around Earth in the ionosphere and the Van Allen radiation belts, particularly near the South Atlantic Anomaly. Both of those forms of radiation behave in somewhat predictable patterns. Scientists know that cosmic rays penetrate the solar system better when the Sun is at a minimum of activity, and some of the patterns of activity in the radiation belts can be explained by daily variations due to the Earth's rotation.

Solar proton events are significantly harder to predict. First of all, not every sunspot group produces a flare or coronal mass ejection, and neither event is necessarily tied to sunspots, so they are extremely hard to predict with any accuracy. Furthermore, not every flare or CME produces high-energy solar protons, and even when they do, those protons may not be directed at Earth. Even when observers manage to spot a proton-producing event on the Sun, it is difficult to determine the magnitude and intensity of the event until it is well under way, so that astronauts will have at best 10 to 100 minutes to take cover from the initial burst. Combine this unpredictability with the damage that can be caused by high-energy protons and you have perhaps the most dangerous space radiation hazard to astronauts and lightly shielded spacecraft.

The penetration of high-energy particles into living human cells and tissues can lead to burns, chromosome damage, cell death, and sometimes cancer. The potency of solar protons arises from what is called ionizing radiation—the proton, or ion, carries enough energy to eject an electron from an atom. When the energy from ionizing radiation is deposited in the human body, chemical changes can occur at the atomic level. The water in the body tends to absorb a large portion of the radiation, and that water can itself become ionized into highly reactive molecules called free radicals. Free radicals can react with and damage human DNA. Alternatively, radiation can collide directly with DNA molecules, damaging them directly.

The most immediate effect of severe ionizing radiation is radiation sickness, a phenomenon that has been studied in depth since the Hiroshima and Nagasaki atomic bomb explosions. The gastrointestinal system, bone marrow, and eyes are perhaps the most sensitive to radiation, and many victims suffer almost immediately from nausea and vomiting after exposure. Intense radiation can also lead to severe burns that are slow to heal, cataracts, and even sterility. The immune system can be severely depleted, with the suppression of immunity allowing infection to overwhelm the body while defenses are down. And while the nerves and brain are most resistant to radiation, acute exposure can damage the central nervous system. Finally, extreme doses of radiation can be fatal within days or weeks.

The slower and more insidious effect of ionizing radiation is genetic mutation. The free radicals created by direct radiation hits can cause unpredictable changes in DNA that can affect both the victim and their offspring (some of the worst effects of radiation likely occur in fetuses in the womb and even within still-to-be-fertilized eggs). Some genetic mutations are relatively harmless, or at least negligible in their effect, though mutations rarely occur for the better. In the worst cases, mutations can disrupt a tissue or cell's ability to control its own reproduction, leading to the uncontrolled growth and division of cells that we know as cancer.

"Any radiation exposure results in some risk," says Mike Golightly, who leads the Space Radiation Analysis Group at NASA's Johnson Space Flight Center, the team that monitors space weather activity during space shuttle and space station missions. "The increase in cancer risk is the principal concern for astronaut exposure to space radiation." With this in mind, NASA and other space agencies set annual and career limits on the amount of radiation an astronaut may endure while traveling and working in space. In fact, the Occupational Safety and Health Administration officially classifies astronauts as "radiation workers." The limit for astronaut exposure to radiation is 25 rem within a 30-day period and 50 rem for an entire year, compared to the limit of 5 rem in a year for people who work with radiation on Earth and 10 cumulative rem over a five-year period. Above 450 rem is considered a median lethal dose—an amount at which 50 percent of victims would die. In terms of harmful radiation exposure, the worst of the Apollo missions—Apollo 14—exposed the astronauts to just

1.14 rem of absorbed radiation. The astronauts who lived on Skylab two years later received a dose of 17.8 rem, and the *Mir* cosmonauts are thought to have received much higher does due to the length of their stay in space. Researchers in the United States have estimated that a one-year stay on Russia's *Mir* space station provided a cosmonaut with 21.6 rem, increasing the lifetime risk of cancer by 1 percent.

When compared with the radiation we receive in everyday life on Earth, the amount astronauts receive seems reasonable for such dangerous work. A simple chest X ray gives a patient 0.01 rem, while the natural background radiation that every human receives from rocks and minerals in Earth's crust is 0.1 rem per year. A dose of 0.1 rem corresponds to an increased chance of 1 in 17,000 of contracting a cancer from such a radiation exposure (compared with the normal incidence of cancer, which is 57 cases per 17,000). The researchers and managers who run the human space flight programs at NASA and other space agencies consider a 3 percent higher risk of cancer—which is still below the risk taken in some earthbound occupations in agriculture and construction—to be acceptable for its flight teams. It's also a fairly low risk when you consider that the average Earth dweller increases his risk of fatality by 1 percent simply by commuting in a car to work each day. And it is a risk that astronauts knowingly take, inasmuch as they are briefed before every mission about space radiation, how much of a dose they can expect for any given flight, and where they stand with regard to total lifetime exposure to radiation.

However, Golightly notes that some research suggests that radiation encountered in space may be more effective at causing biological damage than the gamma rays and X rays encountered by workers on Earth. Moreover, most of the projections and calculations of radiation risk in space are best guesses and theoretical models. Very little is actually known about the biological effects of low-level radiation exposure in space. The astronaut generation is still relatively young and most astronauts are physically robust and healthy compared to the average citizen, so it remains to be seen whether astronauts are more likely to develop cancer than the rest of the population. Furthermore, the number of astronauts past and present is still too small to provide a useful statistical sample to compare with cancer rates in the normal population. Nonetheless, Golightly and his colleagues at the Johnson Space Center note that at least one astronaut has told them that he felt like "a walking time bomb" because of his exposure to space radiation.

NASA astronauts had their closest brush with radiation in October 1989, when a large, long-lived X-class flare produced a swarm of protons that lasted nearly six days. The crew of the space shuttle *Atlantis* was aloft and hard at work for the duration of the October 1989 storm. Though flying well within the protective magnetosphere and at relatively low latitude (away from the auroral zones and the footprints of the radiation belts), the astronauts reported burning in their eyes, a reaction of their retinas to the solar particles. The crew was ordered to go to the "storm shelter" in the farthest interior of the shuttle, the most shielded position. But even when hunkered down inside the spacecraft, some astronauts reported seeing flashes of light even with their eyes closed. On Russia's *Mir* space station, cosmonauts received a significant increase in radiation dose during the solar particle events. According to Badhwar, the total increase in cosmonaut exposure was about 6 or 7 rem, a dose equivalent to 100 to 150 days of additional radiation exposure. And one NASA researcher estimated that there was a 10 percent chance that astronauts on a deep-space mission beyond the magnetosphere or working on the Moon would have died during the October 1989 event.

The close calls may come closer and more often over the next few years. The United States, Russia, and their partners began construction of the International Space Station (ISS) *Alpha* in 1998, and the latest plans call for the work to continue through 2004. More than 40 astronauts are expected to fly 33 space shuttle missions and 10 Russian rocket flights and to work in 6-hour shifts for more than 1,500 hours outside the orbiting station—known as extravehicular activity, or EVA. Much of that activity will occur during the Sun's most violent years, around the peak of the solar cycle. "If you're timing space station construction and the solar cycle, you couldn't have done a worse job," said Mark Weyland, a project manager at Lockheed Martin who studies the effects of radiation with Golightly.

The poor timing is exacerbated by a decision made in 1993 to fly the space station in an orbit that is more accessible from Russia. Under normal conditions, a space shuttle flying below 45 degrees latitude is almost completely shielded from solar flare protons. During large solar events and magnetic storms, the pressure on the magnetosphere and the interconnection of the Earth's field with the interplanetary magnetic field can cause the magnetosphere to be compressed and can allow solar flare particles, trapped radiation near Earth, and cosmic rays to reach latitudes that are typically safe. Since the space shuttle rarely flies higher than 42 degrees—about the latitude of New York City—and since the largest storms from the Sun occur only a few times per solar cycle, shuttle flights are usually pretty safe from space weather. However, the space station *Alpha* flies in a much higher inclination, drifting as far north as 51.6 degrees. That orbit regularly takes the station near the auroral zone and the "horns" of the outer radiation belt—where the trapped particles descend toward the atmosphere. The space station orbit makes astronauts much more vulnerable to solar protons and to other accelerated particles during the largest solar storms.

In 1999 a panel convened by NASA and the National Research Council (NRC) reviewed the radiation risks associated with construction of the space station and criticized the space agencies for their planning and lack thereof in some cases. The group of 10 scientists predicted there is a 100 percent chance that at least two of the 43 missions to the space station will overlap with a solar proton event and a 50 percent chance that at least five missions will be affected. While these space radiation events may not be immediately life threatening, the panel noted, astronauts could be exposed to unacceptable doses of radiation that could put them in danger of exceeding short-term limits for overall exposure and increase their risk of cancer. Even if the health of the astronauts is not affected by one event, the exposure limits could require that missions be interrupted, which would throw off the schedule for future flights and could lead to problems in the rotation of astronauts. Plus it is also a career issue. "No astronaut wants to reach the short-term radiation limits, much less the career limit" because it could end their time as an active astronaut, the panel noted. "The results would seem to call for an aggressive program aimed at reducing solar radiation risk to astronauts during ISS construction."

Preventing the exposure to severe space weather should be simple, as NASA and NOAA both have fleets of spacecraft, ground observatories, and computer models that monitor and predict space weather. If a crew were working outside the station, mission controllers would have about 30 minutes to an hour to get the astronauts inside. While there is no officially designated radiation shelter on the space station, Badhwar noted, "we are mapping the areas using a proportional counter and we have a generally good idea from the shielding distributions—developed from computer-aided design drawings of the ISS modules—to advise the crew of 'safe' areas." So if a particularly egregious solar event is under way, NASA could warn the crew to stick to the most shielded areas to wait out the storm (the first station crew was advised to do so for parts of a solar storm from November 12 to 14, 2000).

But according to George Siscoe, a Boston University physicist and chair of the NRC panel, "an unofficial NASA flight rule specifies that changes in flight plans must be based on current data that reflect the weather immediately around the space station. Information about the size and shape of a solar storm and data on its occurrence, intensity, and duration can be retrieved from other sources, but under current guidelines, this information could not be used by flight directors to take immediate action." Essentially, the flight director for a manned space mission is only supposed to consider real-time conditions in the immediate vicinity of the spacecraft, ignoring the possibility that models and observations might say that solar protons or excited radiation belt particles might be on their way. "These rules unnecessarily restrain ground-based flight directors because other valid data could be used to assist in avoiding radiation exposure," Siscoe says.

Among other things, the panel recommended that NASA install dosimeters on the outside of the space station to provide real-time information about the amount of radiation the crew is receiving while working outside. It also recommended that flight directors ignore the "real-time, onsite" rule and instead rely on the Space Radiation Analysis Group and NOAA's Space Environment Center to provide specific information about the incoming space weather and its potential effects. Finally, NASA, NOAA, and other groups should convene a meeting and find some funding to gather more and better data and to develop more useful, health-specific models of the radiation environment.

"It's not a matter of *if* radiation enhancements will occur while crews are aboard the International Space Station, but *when* and *how serious*," says Mike Golightly. "During construction of the ISS, there is a reasonably large probability that extravehicular activity will coincide with radiation enhancement."

Long after the space station is complete, storms from the Sun could hamper more advanced space exploration efforts. Future missions to set up stations on the Moon or to explore Mars and the outer planets will be impossible until researchers can figure out a way to protect the crew from all of the harmful radiation flying about the solar system. On shuttle flights and many of the manned space missions of the 1960s and 1970s, the astronauts have spent most of their time tucked inside Earth's protective magnetosphere. Though space weather events and severe radiation exposure is possible near Earth, they are almost always mild. But outside of Earth's magnetic field, there is no natural shield from cosmic rays and solar protons. The sheer length of the trip to Mars and back—anywhere from two to three years—ensures that astronauts would face at least one major space weather event. The more insidious threat comes from the long-term, low-level doses of radiation that the crew would take every day for several years. Would the biologically useful bacteria in the human gastrointestinal tract be destroyed? Would the central nervous system be affected, as in clinical studies where long exposures to radiation have affected the function of nerves in lab rats? No one can venture more than a hypothetical guess about the effects of living in unshielded space for so long. "The astronauts are going to have to stay on the Martian surface for as long as a year-and-a-half," Badhwar said, "so they must be healthy or the mission is in real trouble." So as engineers, architects, and doctors figure out how to fly the crew to Mars and how to keep them well fed and mentally stable, physicists must figure out how to keep them from being irradiated to death.³

During that October 1989 solar proton event that sent the shuttle astronauts into hiding within the spacecraft, special sensors that measure cosmic radiation were triggered on a Concorde supersonic airliner. Though the airplane flies much lower in the atmosphere and well below the allegedly radioactive parts of Earth's space, the passengers received the equivalent of a chest X ray from the solar protons. It was an unusual event, as the monitors on the Concorde have rarely been triggered. But it was enough to create interest in the health risks for aircraft crews and passengers who often fly on routes that approach Earth's geographic poles. As with the space station, flights that take polar routes risk greater exposure to particles and radiation from space.

In 1978 a study of the biological effects of air travel on the U.S. population suggested that the exposure to cosmic rays and space weather would be so small as to not be directly observable. Researchers calculated that it was likely that there would be 3 to 75 cases of genetic defects and 9 to 47 cases of cancer over a period of years. In a population of 250 million people, those numbers are so low as to be considered negligible, lower than the number of people who will be struck by lightning. On the other hand, in 12 medical research studies conducted over the past 3 decades, 6 studies found that the risk of cancer was higher for frequent airline flyers (the other 6 studies were inconclusive).

What all of the studies do show is that space radiation is more of a threat to pregnant women. The Federal Aviation Administration (FAA) estimates that the risk of birth defects among pregnant women who fly ranges from 1 in 680 to 1 in 20,000, depending on the frequency of travel and routes flown. Still, that rate is 0.5 percent higher than the rate of birth defects in children of women who do not fly during pregnancy, according to Dr. Donald Hudson, an aviation medicine advisor for the Air Line Pilots Association. "The risk for cancer from low-dose radiation is very low except for female crew members of childbearing age," Hudson notes. "But unfortunately that risk to the fetus is most severe in the first trimester, when women often don't know yet that they are pregnant."

In Europe, where flights to higher latitudes occur more frequently, a law of the European Union went into effect in 2000 that requires all European airlines to educate flight crews about radiation issues. In the United States, the FAA published a directive in 1994 stating: "Air carrier crew members are occupationally exposed to low doses of ionizing radiation from cosmic radiation and from air shipments of radioactive material. . . . It is recommended that workers occupationally exposed to ionizing radiation receive exposure (to the issue) and appropriate radiation practices." Yet very few commercial airlines in North America warn their pilots and flight attendants—much less the passengers—about the issue. The cynical view is that since so few people are aware there is any radiation risk, and no legal judgments citing "radiation exposure" have been made against the airlines, it is not in the industry's interest to address the risks.

Dr. Robert Barish, a medical physicist who has studied in-flight radiation, notes that an airline passenger's exposure to cosmic radiation doubles with every 6,500 feet of altitude, and solar flares can increase radiation exposure by 10 to 20 times. On a typical round-trip flight from New York to Hong Kong—which takes a route closer to the poles and auroral zones of Earth—passengers would receive a radiation dose of 20 millirem (about two chest X rays). The recommended yearly dose of radiation for a person on the ground is 100 millirem (radiation workers are permitted 5,000 millirem [5 rem] per year in the United States, 2,000 in Europe). So after just five flights to Asia from New York, a passenger has already reached the yearly radiation limit without accounting for medical X rays or any other exposures he or she might accumulate.

In his book on the radiation risks of flying, Barish notes that the rates of several types of cancer among pilots "are high—in some cases much higher than average. Radiation levels in a jetliner are occasionally so high that if it was a nuclear power plant, the levels would require signs warning employees not to spend any more time in the area than necessary to do their jobs." Statistics show that the radiation exposure among airline crews is far from lethal, and in most cases the exposure does not exceed the occupational limit. "But their bodies certainly have to handle more than the ground limit exposure for the general public," Barish said. While radiation risk is a relatively minor issue now, Dr. Hudson notes that it will be a much more important issue in the future, when high-speed planes are expected to fly at much higher altitudes

"We have thousands of flight attendants and pilots who receive a radiation dose in the top 5 percent of allowable radiation exposures," says Barish, "yet they are not even counted as radiation workers." On top of that, close to half a million people fly more than 75,000 miles per year, putting them into the same exposure levels as many of the flight crews.

Barish is by no means an alarmist. He just wants the public, or at least airline crews, to be given the right of informed consent. "There is no demonstrable harm at the levels of radiation received by airline crews and passengers," he notes. Most people would not give up flying due to such a minimal risk. "But since environmental and nuclear regulators presume there is a risk, why don't the airlines? The risks are small, but they are there. In every other area where people are exposed to radiation, they are informed. People are entitled to know about that risk."

10 Seasons of the Sun

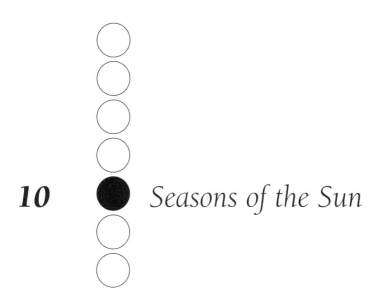

The Sun, with all those planets revolving around it,
and depending on it, can still ripen a bunch of grapes
as though it had nothing else in the Universe to do.
      Attributed to Galileo Galilei

As with the weather on Earth, the Sun and space weather have seasons. More precisely, they have cycles. Close observation of the spots on the Sun or of the intensity of the X rays and ultraviolet light emitted by our star reveals that over the course of about 11 years the Sun's activity waxes and wanes. In response, the Earth's magnetic field and atmosphere become more and less disturbed. Scientists have known the pattern for 160 years, but they remain puzzled by what causes the rise and fall of solar activity. And they are only beginning to address a much deeper question: do solar variability and space weather affect the climate on Earth?

Heinrich Schwabe was the first scientist to observe a cycle at work on the Sun. After observing the Sun daily from 1826 to 1843 and recording the sunspots on its face, Schwabe found that the number of spots per month and per year rose and fell, then rose and started to fall again. After he published his results, Edward Sabine and other scientists compared the rise and fall of sunspot numbers with the frequency of magnetic storms around Earth. They found that the number of storms tracked closely with the number of spots, with more storms occurring when the Sun's face was freckled.

By tracking the Sun methodically for the past 175 years—and extending the observations back another 120 years by compiling and merging disparate records—solar scientists have charted the solar cycle of activity back to the early 1700s. The records confirm that the sunspot cycle lasts about 11 years from minimal activity to a peak or maximum and back to the minimum. The period marked by many sunspots—as many as 200 spots per month, usually about 120 to 130—is known as solar maximum and the quiet period— with fewer than a dozen sunspots or so—is called solar minimum. Some historical solar cycles have lasted as long as 14 years while the shortest ones have risen and fallen in just 9 years (see Figure 15). With the development of better scientific instrumentation, scientists have also detected a 22-year cycle at work, in which the magnetic poles of the Sun flip by 180 degrees and then back again every 22 years. That is, from one solar minimum to solar maximum, the

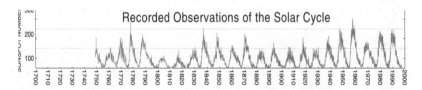

FIGURE 15. Upon close study of the two-and-a-half centuries of sunspot counts, several patterns emerge. It appears that odd-numbered solar cycles tend to be somewhat more intense than even-numbered cycles. And in the past 100 years, solar activity seems to be intensifying compared to the previous century. Courtesy of NOAA-SEC.

Sun's magnetic field flips from a north–south orientation to south–north and then back again in the next sunspot cycle.

Statistical studies of the sunspot cycle also reveal other quirks and oddities that scientists find difficult to explain. For instance, stronger solar cycles (those with more spots) tend to be shorter while weaker cycles tend to last longer. Odd-numbered solar cycles (we are now in the midst of cycle 23) tend to be stronger than even-numbered ones. As the number of sunspots increases, so does the frequency of the Sun's explosions, but not necessarily the strength. That is, the most intense and effective sunspots, solar flares, coronal mass ejections (CMEs), and magnetic storms do not necessarily occur at solar maximum, but instead are spread throughout the cycles. In fact, contrary to intuition, storms tend to be more extreme after the peak of the solar cycle, when the Sun is allegedly calming down and returning to solar minimum.

Another statistical quirk is that each solar cycle usually has three different peaks: one for the number of sunspots, one for the number of flares, and one for the number of geomagnetic storms. And while flares and coronal mass ejections are much more common at solar maximum, coronal holes and the damaging high-speed solar wind streams they produce are much more common during the approach to solar minimum. The failures of the Anik (1994) and Telstar 401 (1997) satellites as well as the huge solar flares of 1972 all occurred as the Sun was muddling through the middle, minimum periods of its activity cycle.

The progress of sunspots as they move around the face of the Sun may provide a hint of what is happening inside the star. Years after his discovery of solar flares in 1859, British scientist Richard Carrington completed a study of the position of sunspots throughout the solar cycles. He found that sunspots appear closer and closer to the equator as the Sun becomes most active. At solar minimum, the spots tend to appear around 30 to 40 degrees of solar latitude and as the Sun progresses toward the maximum, the spots develop at lower latitudes, nearer to the middle of the Sun.

According to a theory proposed in the 1960s by astrophysicist Horace Babcock, the sunspot cycle and the migration of sunspots toward the equator is likely a result of the differential rotation of the Sun, whereby the equator spins faster than the poles. The Sun starts with a relatively simple north–south set of magnetic poles. But as the midsection of the Sun spins faster than the higher latitudes—in three years the equator would lap the north and south Poles more than five times—the magnetic field of the Sun gets twisted and pulled in the east–west direction. This twisting and shearing of the magnetic field lines creates two doughnut-shaped "toroidal" fields, one in the north and one in the south. The intensity of these snarled magnetic fields causes them to erupt out of the Sun and create active regions of sunspots. As the solar cycle progresses and the equator continues lapping the poles, the toroidal fields move down toward the equator, carrying their sunspots with them. Eventually the two fields cancel each other, while the process starts all over again at higher latitudes.

Based on several centuries of sunspot watching and the predictions of numerous solar physicists, the process of magnetic twisting and snarling inside the Sun was expected to reach a crescendo in 2000 through 2002. Solar cycle 23—the twenty-third since reliable measurements were first made in the early 1700s—began in October 1996, when the Sun started its ascent out of solar minimum. In September of that year, the National Oceanic and Atmospheric Administration (NOAA) and the National Aeronauticcs and Space Administration (NASA) convened the Solar Cycle 23 Project, an international panel of scientists who were charged with making the best scientific guess at the magnitude of the current solar maximum. Scientists from around the world submitted 28 predictions based on six different prediction methods. Groups studied the relationships between the coming solar cycle and the length of the previous cycle, the level of activity at sunspot minimum, or the size of the previous cycle. Some scientists analyzed measurements of changes in the Earth's magnetic field at sunspot minimum, which seem to have a statistical relationship to the intensity of the next solar cycle. Still other groups examined the coronal holes and the strength of the solar magnetic field at solar minimum to say something about Babcock's theory of magnetic turmoil inside the Sun.

But in the end, since scientists are only starting to view the inside of the Sun—and they still don't know what to look for—the predictions are ultimately educated statistical guesses. To David Hathaway, a solar physicist at NASA's Marshall Space Flight Center and a contributor to the Solar Cycle 23 Project, the solar cycle prognostication is akin to the weather predictions in the *Farmer's Almanac*. "It's like saying we're going to have a mild or cold winter," he notes. "In the end, it's all statistical inferences. There's no real physics involved."

After the Solar Cycle panel compared and analyzed all of the predictions, it was decided that solar maximum 23 would most likely occur in March or April 2000 but could happen as late as January 2001. The panel predicted that the maximum number of sunspots would be about 160, though the range was listed as anywhere from 130 to 190 sunspots. The average solar cycle has a maximum monthly sunspot number of 110. In effect, the new solar cycle should look a lot like the last one, which peaked in July 1989 as the third-largest solar cycle ever observed.[1]

The Sun has teased scientists as it has meandered through solar maximum 23. The violent sequence of CMEs and flares in May 1998 seemed to announce the approach of a boisterous solar maximum. But then the Sun bobbed and weaved its way up the slope of the predicted sunspot curve (see Figure 16). The monthly sunspot number rose sharply to 137 in June 1999 but dropped to 70 by September. The Sun really picked up its pace in March 2000, reaching a sunspot count of 138 and sustaining active levels through the summer. In July 2000 the sunspot number reached 169, the highest number of the current solar cycle. Living up to the title of "solar maximum," July 2000 brought the Bastille Day storm, a monstrous space weather event that killed at least one satellite and brought the auroral oval down to the Gulf Coast of the United States.

By the autumn of 2000, the Sun cooled off and less experienced Sun watchers began to speculate that solar max had come and gone. But the solar roller coaster began another steep ascent in March and April 2001, reaching 134 sunspots. More than a

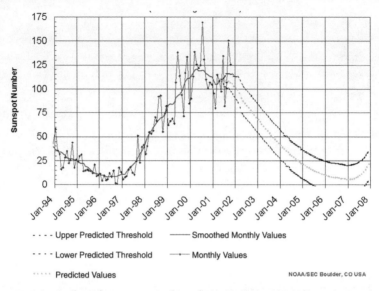

FIGURE 16. A plot of sunspot numbers from January 1994 through October 2001 shows the erratic decline and rise of solar activity during solar cycle 23. The plot includes the projections made by the scientists of the Solar Cycle 23 Project. Courtesy of ISES/NOAA-SEC.

year after the supposed "maximum" the Sun was covered with 150 spots in September 2001. The ride seemed like it wouldn't end.

Ever since space weather researchers began their media campaign to educate the public about solar maximum, people have been questioning scientists and marking their calendars for the *day* or *month* of solar maximum. They continually ask, "Have we reached the maximum?" But in reality the peak of the solar cycle is just the top of a statistical curve. One month may mark the absolute highest point in the cycle, but the Sun roils with maximum activity for years on either side of that historic month. "The sunspot maximum is usually a broad peak," said Hathaway. "There is a two- or three-year period when activity is quite high." So while the statistics say that the Sun was at its most boisterous in July 2000, scientists wouldn't be surprised if the actual surge of solar activity lasted into 2003.

"On some previous cycles, there was some stagnation and hesitation by the Sun on the way to solar maximum—a plateau in the middle of the rise," JoAnn Joselyn, a researcher at NOAA's Space Environment Center (SEC) and chair of the Solar Cycle 23 Project. "This cycle is above average in terms of solar flux."

SEC, which is responsible for monitoring and predicting space weather for the U.S. government, spent much of the late 1990s warning the public and its commercial customers to expect solar proton radiation showers to become more frequent and more intense during the period of solar maximum. The center made its own prediction based principally on the experiences of the past few solar maxima, that the *aurora borealis* (northern lights) would make several appearances over the continental United States in the next few years, reaching as far south as the Gulf of Mexico at least once. The SEC team wrote that two magnetic storms of the magnitude of the March 1989 storm were possible and the "probability for severe geomagnetic storms will be the greatest during an extended period lasting from 1999 through 2005."

Even if the Sun does not follow a scientist's schedule, even if the storms from the Sun during cycle 23 are no worse than any in the past, the current solar maximum and the next one in 2011 are expected to have a much greater impact on society than any we have endured before. Two to three times more satellites are flying now than during the maximum of 1989, and civilization relies on those spacecraft for a lot more information and communications. Electric power grids are serving more customers, but with about the same number of transformers, power plants, and transmission lines as were in operation 11 years ago. "The explosion in technology is intersecting with an extremely disturbed space environment," said Joselyn. "Electricity is no longer a luxury, and satellites are becoming a critical link in society. There is much higher risk now because we depend more on technology that is vulnerable to space weather."

Though their lives and careers are separated by four generations, E. Walter Maunder and Jack Eddy are linked intellectually and spiritually by the Sun. At separate times over the past 100 years—Maunder in the 1880s and 1890s, Eddy in the 1970s—the two solar researchers dared to assert what many of their colleagues would not. The clockwork 11-year solar cycle is not constant, and it is probably not the only cycle at work in our star. Over geologic timescales, other cycles and patterns may be unfolding. Studying historical records and bucking conventional wisdom, Maunder and Eddy saw a connection between the Sun and Earth that scientists are cautious to believe but hard pressed to dismiss.

Intrigued by the observations and conjectures from eighteenth-century astronomer William Herschel and nineteenth-century schoolteacher-turned-astronomer Gustav Spörer, Maunder began investigating a phenomenon he would one day call "the prolonged sunspot minimum." Spörer and Herschel both noted that sunspots seemed to disappear from the astronomical records for nearly 70 years in the late seventeenth and early eighteenth centuries. As the superintendent of the solar division of Britain's Royal Greenwich Observatory in the 1880s and 1890s, E. W. Maunder had access to hundreds of years of logs, journals, and historical records. So he began to dig deeper into this disappearance of sunspots. His studies confirmed that from about 1645 to 1715, fewer spots were seen on the Sun over those seven decades than can typically be seen in a single active year. Most of the spots that did appear were located near the solar equator and scarcely lasted for more than one rotation of the Sun. From 1672 to 1704, no spots were observed at all in the northern hemisphere of the Sun. And from 1645 through 1705, no more than one sunspot group showed up at a time, and often there was just one lonely spot rather than a group.

Maunder presented his findings and a summary of Spörer's work to the Royal Astronomical Society in 1890 and again in 1894, titling his papers "A Prolonged Sunspot Minimum." He argued that the period of low sunspot activity would eventually prove important to understanding the Sun and how it affects Earth. But apparently his work fell on deaf ears, because little mention of it is made in the science publications of the early twentieth century, and Maunder himself felt compelled to reprise the story with an updated version in 1922. His later work incorporated observations suggesting that very few auroras were observed during the time of the quiet Sun.

Perhaps the initial thought by most scientists was that astronomers had simply missed the sunspots during the so-called minimal period, having lost interest after the initial buzz of Galileo and Christopher Scheiner in the early 1600s. But the historical evidence from the period contradicts that notion. In the latter half of the seventeenth century, scientists such as Johannes Hevelius and Jean Picard actively chronicled solar activity. And there were certainly other great astronomical discoveries made during the period—including the first observations of Saturn and its rings and moons, the observation and study of Halley's comet, and the calculation of the speed of light from observations of Jupiter's moons. Moreover, the few sunspots that were observed during the period were usually an occasion for scientific notice and publication. Astronomers Giovanni Cassini and John Flamsteed both paid special attention to sunspots they observed in 1671 and 1684, respectively, noting that such sightings had been rare in recent years.

For whatever reason, Maunder's sunspot minimum did not capture much attention until American solar physicist Jack Eddy revived the studies in the mid-1970s. Provoked by stories that University of Chicago space physicist Gene Parker told him about Maunder, Eddy reviewed the turn-of-the-century papers. His initial intention was to debunk the theories about solar variability and climate. "I had been taught that while the Sun indeed affects the upper and outer atmosphere of the Earth, purported connections with the troposphere and weather and climate were uniformly wacky and to be distrusted. . . . The claims that were made for associations between weather events and the Sun I thought were pretty preposterous. The trail was, initially, purely historical and driven by my prejudice of trying to find examples from the past that would disprove, once and for all, the notion of strong Sun-weather relations. A devout negativism on this subject was the gospel at the

High Altitude Observatory, where I had been trained. . . . It needed to be shot at, even after all these years, and dismissed once and for all. So I set out to demonstrate that what Maunder had claimed was really nonsense. . . . I was trying to examine the early origins of Sun-weather claims, like unrolling and deciphering the Dead Sea Scrolls of solar physics. But it was mostly a love of history that took me down the trail."

To his surprise, Eddy gradually learned that the observations of Spörer and Maunder were anything but nonsense. He expanded on their work by pulling in historical records of auroras, naked-eye sunspots, and eclipses. According to one catalog that Eddy cited, only 77 auroral shows were recorded in the entire world from 1645 to 1715, and 20 of those appeared in 1707 and 1708, when sunspots were present. Of the aurora reports that were made, nearly all of them originated from Scandinavia, despite the fact that London and other locations in Europe typically see 5 to 10 northern light shows per year. Eddy could not find any reports of sunspots viewed with the naked eye in China or other Asian countries, where naked-eye sunspots had been recorded as momentous events since antiquity. The historical records from the Far East typically cite at least one naked-eye sunspot per decade.

Eddy also noted that while astronomy was budding and scientists were chronicling everything they could see in the sky, not one of the accounts of eclipses during the period of Maunder's sunspot minimum describes the corona of the Sun in any detail. In the usual eclipse, the streamers and detailed structures of the solar wind can be seen extending for some distance as the Moon blocks the dazzling disk of the Sun. Yet at a time when sky watchers would have been more attuned to celestial phenomena, no one was writing about the solar corona or what it looked like through a telescope.

But the clinching evidence for the "Maunder Minimum," as Eddy came to call it, could be found in the growth rings of fallen trees. As trees lay down their rings each year, they record information about the atmosphere and the environment. Specifically, trees capture carbon dioxide, which consists of a blend of two carbon isotopes—

carbon 12 (C^{12}), which is chemically stable, and carbon 14 (C^{14}), which is radioactive. C^{14} is constantly being formed in the Earth's atmosphere when cosmic rays bombard the carbon and nitrogen compounds in the air. That rate of C^{14} formation varies ever so slightly, but the rate at which the isotope decays is constant (the "half-life," or time it takes for one-half of the C^{14} to decay into nitrogen, is about 5,600 years). When the Sun is more active, fewer cosmic rays make it into Earth's atmosphere, producing less C^{14}; when solar activity is low, cosmic rays flow unimpeded to Earth and create slightly more C^{14} for trees to absorb.

Combining the records of the solar community with those of environmental scientists, Eddy found that tree rings from the era of the Maunder Minimum contained significantly more C^{14} than the trees and tree rings formed in the years before or after the sunspot minimum. When held up next to the historical records of auroras and sunspots—and their absence from 1645 to 1715—it became obvious to Eddy that something in the Sun-Earth relationship had changed for those 70 years. The constant solar cycle was in fact not constant at all.

What makes the Maunder Minimum more than just a curiosity of solar science are the stark changes in Earth's climate that seemed to coincide with the lack of sunspots. In the decades of Maunder's sunspot disappearing act, Europe and North America endured a period of extremely harsh winters and colder than normal summers; the period has been dubbed the Little Ice Age. Glaciers advanced farther south than they had at any time since the end of the last true Ice Age (about 15,000 years ago). No major floods washed over the lowlands of Switzerland—suggesting that alpine snows and glaciers hardly thawed as they normally do with the rise and fall of the seasons. The average global temperature dropped one degree below the norm, but that was enough of a fluctuation to cause severe drought in the American Southwest and to allow Londoners to skate on the Thames.

Studies of older trees and their rings have since revealed other periods when the Sun appears to have waned and waxed beyond the norms of the 11-year cycle. According to the C-14 record, the

Sun activity dropped off noticeably from 1460 to 1550, another period of global cooling now known as the Spörer minimum. Conversely, from about 1100 to 1250, the Sun waxed into a prolonged maximum of activity, a period that coincides with the "Medieval Climatic Optimum." Temperatures apparently rose enough to allow the Norse people to first settle Greenland and give it the name that now seems so preposterous. The period also induced a severe, decades-long drought that may have brought on the demise of the Anasazi culture of North America.

While it is tempting to immediately connect the Maunder and Spörer minima and the medieval maximum to changes in Earth's climate, scientists have remained skeptical for much of the past century. Sure, there have been changes in climate that coincide with changes on the Sun. The Maunder Minimum coincided with a cooling period in many parts of the Earth, and in the 300 years since then, solar activity has steadily increased at the same time as global mean temperatures have increased.

But many of the scientists who run computer models of the atmosphere do not seem to have room for the Sun, or at least for one that varies. And many geophysicists are wary of the solar input because they cannot explain the physics and chemistry of how solar variability could change the climate. The question remains: is the concurrent rise of solar activity and of earthly temperatures a coincidence or a cause-and-effect relationship?

The Sun dominates and drives climate patterns on Earth. Depending on the season of year and the tilt of the Earth's axis toward or away from the Sun, temperatures climb and fall by hemisphere. The southern hemisphere warms and the northern hemisphere cools as the axis tilts away from the Sun. In summertime, sunlight beats down directly over one hemisphere and the beams fall more obliquely over the other hemisphere. The tilt of the axis causes one-half of the planet to bathe in longer periods of warming daylight than the other. Effectively, the amount and intensity of sunlight pouring out from the Sun to Earth do not change, but the angle of Earth's axis forces more sunlight to shine longer over one region than the other. This sunlight warms not only the surface of the planet but also the atmosphere and the oceans, which both retain heat and drive weather patterns. It is a cyclical—seasonal—pattern, but one that appears quite stable.

Until recently, the radiative output of the Sun was generally thought to be constant—that is, scientists assumed that the Sun emits a steady and unchanging amount of light and heat. But, in fact, the amount of radiation from the Sun does vary. While it would seem that all the spots on the Sun would dim the light a bit, in fact the Sun is brighter during solar maximum because there are more bright active regions. Measurements from the Active Cavity Radiometer Irradiance Monitor (ACRIM) experiment on NASA's Solar Maximum Mission satellite and the Earth Radiation Budget (ERB) experiment on Nimbus-7 revealed that the Sun's output changed by 0.1 percent from solar maxima to solar minima in the 1970s through early 1990s.

While the total amount and intensity of sunlight—total solar irradiance—varies by 0.1 percent, other parts of the solar spectrum undergo more significant changes. In particular, high-energy, short-wavelength forms of light are more intense during the Sun's most active periods, with ultraviolet (UV) radiation increasing by 6 to 8 percent. UV rays don't contribute much to the total radiative output of the Sun, but they are one of the major contributors to the cyclical changes in irradiance. And since ultraviolet radiation is known to be an important catalyst of chemical reactions in Earth's atmosphere, small increases in solar UV might have big consequences for the energy balance of the upper atmosphere.

Analysis of the changes in solar irradiance suggest that the variation from solar minimum to maximum could produce a global averaged temperature change of about 0.2 degrees Celsius, or about 0.24 watts per square meter of the Earth's surface. More recent observations from the Upper Atmospheric Research Satellite (UARS) confirm that this cyclical rise and fall is ongoing. A variation of 0.2 degrees seems trivial and almost silly to worry about.

Predictions of global warming suggest that the global surface temperature could rise by 1 to 3 degrees Celsius (2 to 5 degrees Fahrenheit), numbers that seem negligible when one considers the daily fluctuations most of us feel every day. But in a recent editorial, Jack Eddy put such temperature changes into context: "Why should a change of but a few degrees, 50 or 100 years in the future, be of such concern today? For those of us in middle latitudes, the variation from day to night in the temperature of the air can be 20 degrees C, and in the course of the seasons the daily mean varies through 50 to 60 degrees. What is it about a global average that makes a change of 1 degree profound? A change of 1 degree in the mean implies larger changes in parts of the globe—particularly polar regions—but also smaller ones in others. And a small variation in temperature can signal much greater changes in other conditions, such as precipitation and storms and river flow." Eddy notes that global mean temperatures have varied by no more than 1 degree over the past thousand years—a period that included the wicked winters and cool summers of the Little Ice Age. A global increase of just 2 degrees C would double the number of unusually hot days on Earth. And the difference between today and the last major Ice Age—when a mile-thick layer of ice covered most of North America—is only about 5 degrees C in the global mean value.

According to Drew Shindell, a climate researcher at NASA's Goddard Institute for Space Studies in New York, the variations in solar radiation over the course of a solar cycle might affect the ozone layers of the atmosphere. Observations from UARS show that the additional high-energy radiation at solar maximum increases the amount of ozone in the upper atmosphere by at least 1 or 2 percent. The increase in ozone warms the upper atmosphere, and this warm air affects wind patterns from the stratosphere all the way down to the troposphere (the lower layer of the atmosphere near the surface). "When we added the upper atmosphere's chemistry into our climate model, we found that during a solar maximum major climate changes occur in North America," noted Shindell. Specifically, westerly winds blow stronger, but changes in wind speed and direction occur all over the planet. "Solar variability changes the distribution of energy." Over an 11-year solar cycle, the total amount of energy coming into the atmosphere does not change very much. But where that energy gets deposited varies. This leads to changes in wind speeds and prevailing directions, which create different climate patterns.

Other scientists, such as Henrik Svensmark of the Danish Space Research Institute and Brian Tinsley of the University of Texas at Dallas, suspect that changes in the Sun's output might affect the cloud cover of Earth. Cosmic rays, which originate outside the solar system when stars go supernova, bombard the atmosphere and change the chemistry a bit. The reaction produces aerosols that help seed and form clouds. The influx of cosmic rays also can cause more electric charge to accumulate on the ice crystals and evaporating droplets at the tops of clouds. This accumulation of charge can in turn affect how clouds reflect light back into space, how much water falls out as precipitation, and how heat is stored and released in the atmosphere.

The link between cosmic rays, solar activity, and climate seems to be the interplanetary magnetic field (IMF). The IMF is blown out from the Sun by the solar wind and the strength of the magnetic field reflects the intensity of solar activity. The IMF causes the cosmic-ray flux (as measured at Earth) to vary by deflecting these high-energy interstellar particles away from the inner solar system. When the Sun's activity reaches a minimum, the IMF tends to be weaker and more cosmic rays can wash over the Earth's atmosphere. But at solar maximum, the heightened activity of the Sun and the more potent IMF prevents much of that cosmic radiation from reaching the inner parts of the solar system. During a prolonged sunspot minimum—such as the Maunder Minimum—the IMF would likely be very weak, allowing more cosmic rays to reach Earth and producing more clouds that would reflect sunlight. Could this be a causal link between the Maunder Minimum and the Little Ice Age?

Researchers recently stirred up controversy by unearthing a long-term trend from records of solar and space weather activity. Mike Lockwood and colleagues at Britain's Rutherford Appleton

Laboratory have found evidence suggesting that the strength of the solar magnetic field in the IMF has doubled over the past 150 years. They deduced this trend by using a combination of solar wind measurements (dating from 1963 to the present) and Earth-based magnetic field measurements dating back to the mid-1800s. Other researchers such as Ernie Hildner, director of NOAA's Space Environment Center, argue that there is no evidence that the solar magnetic field has been increasing over the last 40 years. The question of whether the IMF has been increasing is not just an esoteric argument. According to the cosmic-ray hypothesis, an increasingly potent IMF would lead to a lower influx of cosmic rays, less cloud cover, and more solar radiation reaching the surface of Earth. An extended period of a weak IMF could lead to greater cloud cover and lower temperatures.

In another connection between space weather and terrestrial weather, there is compelling evidence that dynamic changes in the Earth's stratosphere may be synchronized to the solar cycle. For nearly three decades, Karin Labitzke of the Free University of Berlin and Harry van Loon of the National Center for Atmospheric Research have worked together to show that a known 10- to 12-year oscillation in the stratosphere of the northern and southern hemispheres matches up fairly well with the four solar cycles since 1958. By analyzing average temperatures and air pressures in the stratosphere, they found that the height of the stratosphere rises by as much as one-third as the Sun progresses from solar minimum to maximum. The stretching of the stratosphere likely affects wind and air pressure patterns in the lower atmosphere. The solar cycle effect is not strictly global; in fact, the stratospheric changes are most pronounced near the poles and between 20 and 40 degrees of latitude. But changing the atmosphere in any one part of the world still affects global weather patterns everywhere, as weather changes in one region and forces changes in the neighboring regions.

"We were all taught that the solar cycle had no influence on climate," notes van Loon. "Even in recent years, the 0.1 percent difference in solar irradiance has been written off as 'noise.' But the solar-stratospheric relationship is more than a statistical coincidence. Any relationship between changes in solar output and what happens here on Earth is important for understanding long-term climate."

While the work of van Loon and others correlates the 11-year solar cycle to other 10- to 12-year cycles of activity at work on Earth, a longer view reveals even more Sun-climate connections. Reviewing the past 130 years of sunspot records and temperature measurements, Eigil Friis-Christensen and Knud Lassen of the Danish Meteorological Institute and John Butler of Ireland's Armagh Observatory found that global temperatures seem to fluctuate in synch with the solar cycles. Shorter solar cycles—that is, those that last fewer than 11 years—tend to produce warming trends on Earth, while longer solar cycles coincide with periods of cooling. As the studies of the sunspot numbers and tree rings show, the overall trend of the past 400 years has been toward greater solar activity and higher temperatures.

Recent studies of other stars and computer models of our own star suggest that solar irradiance—the intensity of the light emitted by the Sun—decreased by 0.15 to 0.4 percent during the Maunder Minimum of the seventeenth and eighteenth centuries, four times the variation observed from maximum to minimum in the 1970s and 1980s. Given such a drastic change over a short period of history, and given that we have methodically observed the Sun and the climate for only a few centuries, there is the reasonable possibility that the Sun-Earth system may undergo much more dramatic variations over longer timescales.

The problem now is time and politics. Scientists can measure and model and mull how Earth's climate changes over time, but until they have a longer and more substantial set of data about our environment, most will remain skeptical about the Sun-Earth connection to climate. Researchers will either have to develop a way to project solar activity back into ancient times or they will have to wait until their grandchildren's grandchildren have lived through the next major warming or cooling spell to know for sure. "The Sun's energy variation does affect weather, and it should not be discounted," says van Loon. "But the role of the Sun in climate change is just a big unsolved problem."

The other problem is that some political groups give some Sun-climate researchers an unfortunate and misappropriated reputation. Groups point to the work of scientists like Svensmark or Butler on the solar role in climate change and claim that their work is compelling evidence that human activity has not contributed to the warming of the Earth. Even the most devout Sun-climate researchers acknowledge that it is irresponsible and at odds with the basic physics of the greenhouse effect to say that industrial emissions of carbon dioxide and other gases have not affected the atmosphere. The increase in greenhouse gases is a measurable fact, as are the global rise in temperatures, the depletion of atmospheric ozone, and the increase in ground-level ozone.

On the other hand, it would be equally irresponsible to reflexively claim that *all* of the observed global warming is anthropogenic and that the Sun has nothing to do with it. We have to face the fact that the Sun is the primary driver of our climate system and that we are living with a star that changes.

11 The Forecast

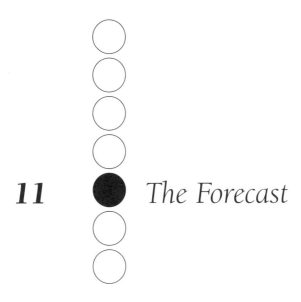

Mama always told me not to look into the sights of the Sun. . . .
Oh, but Mama, that's where the fun is . . .
                    Bruce Springsteen, "Blinded by the Light"*

They sit in front of a bank of computers and monitors, perhaps a dozen glowing phosphorescent screens. They pore over mathematical wiggles and line plots, spacecraft images of the Sun and the aurora, and old-fashioned hand-drawn sketches of the solar surface like the ones Galileo and Carrington made centuries ago. From this fluorescent-lighted and white-walled room tucked into the center of a new office building, they cannot see the Sun, at least not the visible one warming Boulder, Colorado, on a summer day. "They" are the men and women who forecast space weather for the United States and much of the world. They

---

*Copyright © 1972 by Bruce Springsteen (ASCAP). Reprinted by permission.

work for the Space Weather Operations Center—a joint operation of the National Oceanic and Atmospheric Administration's (NOAA) Space Environment Center (SEC) and the U.S. Air Force—as the official monitors of solar activity and of conditions in the space around Earth.

Life as a space weather forecaster in the operations center is hardly glamorous. Unlike the dramatized control rooms of the Apollo missions to the Moon or of Hollywood science flicks, the day-to-day routine in the SEC forecast center seems almost monotonous and uneventful. And most days the forecasters probably prefer it that way. Excitement at the SEC means billions of dollars worth of equipment and commerce are being threatened by an environment we can barely see and are only beginning to understand.

On one morning in June 1999, Eric Ort was watching the Sun and its antics. A lieutenant in the NOAA uniformed corps of officers, Ort was the official forecaster for the day, charged with predicting what the Sun would do next. He scanned the images and radio data from telescopes at solar observatories in Australia, Puerto Rico, Massachusetts, Italy, and New Mexico. Picking a fax out of the machine, he reviewed the hand sketches of sunspots, coronal holes, filaments, and prominences of the Sun as drawn by observers at Holloman Air Force Base in New Mexico. Ort looked up at the many plots of data on the computer screens above and around him: measurements of solar X rays and high-energy particles bombarding the magnetosphere and magnetometer readings of changes in the Earth's magnetic field. He glanced at e-mails and imagery on the World Wide Web from the science team of the Solar and Heliospheric Observatory (SOHO), which takes images of the Sun 24 hours a day. He scanned the real-time data from the Advanced Composition Explorer (ACE) satellite, which gives Ort *in situ* measurements of the speed, magnetic orientation, and density of the solar wind from a position about 1 million miles toward the Sun, in front of Earth.

Marshalling all of his data, plus a few years of experience and training, Ort devised a probability for what the solar activity would be for the next three days and whether that activity would affect Earth. He followed certain algorithms and statistical formulas developed by the research scientists down the hall and made an educated guess about the future. On that day he predicted that the planetary K index (Kp)—a measure of the space weather storm activity around Earth, similar to the Saffir-Simpson scale of hurricane intensity—would be between 1 and 3 on a scale that goes to 9. "It should be a quiet day," with minimal activity, Ort noted as he compiled his forecast, "though there is a coronal hole on the western limb of the Sun that could be geoeffective."

Had this been March 1989 or May 1998, when the Sun was teeming with activity, the prediction would have been much more difficult and probably much less reliable. Ort might have had to issue an "alert" to say that a significant solar event had been observed, a "warning" that a space weather event near Earth was "highly probable," or a "watch" asserting that conditions were favorable for a space weather event. But the Sun seemed to be behaving that day, so he thought he would have some time to catch up on his paperwork and his ongoing education in space science. Based on the outcome of his prediction—the Sun barely peeped for the two days after Ort's prediction and the Kp index never rose above 2—it appeared that his education was coming along nicely.

Electric power companies, surveyors, satellite operators, the military, radio operators, some parts of NASA (such as the Space Radiation Analysis Group at the Johnson Space Flight Center)—even pigeon racers—all look to Ort and his fellow SEC forecasters to warn them of incoming storms from the Sun. The job of space weather forecaster is comparable to the work of the forecasters at NOAA's National Weather Service, according to Ernie Hildner, director of the Space Environment Center. The forecasters must specify the current conditions (Is it raining protons? Is the solar wind blowing hard?), make predictions about possible changes in that environment, and issue warnings to the nation about the space weather equivalent of hurricanes, tornadoes, and blizzards. But unlike their colleagues tracking terrestrial weather—who have a reasonable understanding of how weather moves and changes after nearly two centuries of daily reports—SEC forecasters have to roll the dice daily because the science is still so new, the data are so sparse, and the tools of the trade are still being developed.

"In the beginning, in the 1960s, all we did was observe the Sun with one solar telescope, one riometer, and one magnetometer," notes Gary Heckman, a 35-year veteran of the space weather center and the senior forecaster. "About a decade ago, we got the right day of a major storm about 25 to 40 percent of the time"—when they even knew that a storm was coming. "Now we are about 50 percent correct at predicting the big storms." Most of the staff at SEC will readily admit that the science or art—depending on one's perspective—of predicting space weather is about as mature as terrestrial weather prediction was in the 1960s.

"We need a better set of measurements, more observational tools, and better models" to really become more precise in the predictions, Heckman adds. Using a limited amount of data and a limited understanding of how the Sun generates storms, Heckman, Ort, and others must venture a prediction each day about what the Sun will do for the next three days and how turbulent the space around Earth will be as a result. They do this with timely information from just a handful of satellites and a few dozen ground-based observatories—there are more local weather stations in some states than the SEC has available for monitoring billions of cubic miles of space in three dimensions. They cannot see solar flares or coronal mass ejections from ground stations. And even with the spectacular views of the Sun provided by spacecraft such as the European Space Agency's SOHO, Japan's Yohkoh, and NASA's Transition Region and Coronal Explorer (TRACE), the forecasters and scientists cannot say with much certainty when a solar storm will or won't cross paths with Earth. They can only guess at the direction and trajectory of a storm until it actually washes over ACE, NASA's spacecraft monitoring the solar wind from a position 1 million miles out in front of Earth.

According to Heckman, the introduction of real-time solar wind data from ACE has been the greatest recent improvement in space weather monitoring because it essentially provides a "nowcast." Sitting in a direct beeline from the Sun and bathing constantly in solar particles and radiation, the ACE can detect trouble just before it reaches Earth. This allows forecasters to give warnings about 30 to 60 minutes before a storm arrives. To date, SEC forecasters have been right as much as 80 percent of the time when a disturbance is detected by ACE, according to Heckman. "Once you get past one hour, however, the forecast becomes much less reliable."

An hour of advanced warning about space storms is useful for the electric power companies because they can temporarily change the way they generate and move electricity. Depending on the time of day and the amount of electricity being consumed, companies can shed some of the load on the systems, increase their power-generating capacity, or defer maintenance on some power lines until the threat has diminished. For instance, in a July 2000 storm, the U.S. Nuclear Regulatory Commission advised nuclear power plants to operate at 80 percent of capacity so that they would have a margin of error if the magnetic storm produced strong geo-magnetically induced currents.

But for the satellite operators it can cost too much money and time to adjust to every potential blip in the space environment. If operators are concerned about space weather events during a period when they need to issue a lot of commands, they will sometimes postpone those commands—even a rocket launch—until a storm passes. But rarely do spacecraft get shut down altogether due to a possible storm. Operators instead go into a heightened state of alert, preparing for recovery from a failure rather than trying to prevent one. Space weather failures may cost a lot of money, but so do false alarms when there are too many of them. "If you shut off the satellite, you lose the income stream that the satellite would otherwise be generating," notes Ernie Hildner. "For any single storm, the probability of your satellite getting hit is so low that turning it off is just not the economical thing to do." Right now, the SEC predictions are too variable too often for most satellite companies to spend money mitigating a problem that may never occur. It is a calculated gamble they have to take.

Since the flights of Explorer 1 and Sputnik in the late 1950s, the Sun-Earth system traditionally has been studied as a set of independent parts—the Sun, interplanetary space, the magnetosphere, the ionosphere, and Earth's upper atmosphere. Over the past 40 years, space scientists have developed and refined instruments in support of dozens of individual satellite missions, including the Interplanetary Monitoring Platforms, the Solar Maximum Mission, the Dynamics Explorers, Helios, and the International Sun-Earth Explorers. But each of those programs, while greatly advancing one aspect of the science of solar-terrestrial physics, usually made their measurements in one region of space with one set of instruments and parameters that were not necessarily set up to match what was being studied by other spacecraft. Real-time coordination and correlation with other missions, scientists, and instruments occurred more often by happenstance than intention.

To better understand the Sun, the Earth, and the space in between as an integrated dynamic system, scientists eventually decided that they needed a program of simultaneous space- and ground-based observations coupled with theoretical studies. They had to find a way to assess the production, transfer, storage, and dissipation of energy from one region to the next, with most of that energy moving in ways that are invisible to traditional telescopes and cameras. Most of all, they needed to gather and store data in formats that any space scientist could use. In essence, they needed a comprehensive, quantitative study of the movement of energy from the surface of the Sun to the surface of the Earth. It took two decades of discussions, planning, coordination, and budget fights—particularly because no one nation could afford to pay for such a massive undertaking—but the scientific community came up with an answer: the International Solar-Terrestrial Physics (ISTP) program.

ISTP was conceived in the 1970s, planned in the 1980s, and launched in the 1990s. The mission was intended as a global effort to observe and understand our star *and* its effects on our environment. An armada of more than 25 contributing satellites—working together with ground-based observatories, computer simulators, theoretical modeling centers, and data repositories—allowed scientists to study distinct but connected parts of the Sun-Earth system simultaneously from many perspectives and in many different ways.

The primary missions of ISTP—Geotail, Wind, the Solar and Heliospheric Observatory, Polar, and Cluster II—allowed physicists to observe all the key regions of Earth's neighborhood in space. The first satellite, Geotail, was launched in 1992 by Japan's Institute of Space and Astronautical Science (ISAS) to study the distant reaches of Earth's magnetic tail, the region downwind on the nightside of Earth. In 1994, NASA launched the Wind mission into an orbit that sent it on long loops toward the Sun and around the Moon, allowing the satellite to sample the solar wind from outside the magnetosphere. Late in 1995, the European Space Agency (ESA) and NASA launched SOHO to observe the Sun in several wavelengths, full-time, from space—without the intrusion of Earth's shadow—and to conduct studies of activity inside the Sun. In 1996, NASA added the Polar satellite to swing over Earth's North and South Poles in order to monitor the aurora and the flow of currents and plasmas into and around the magnetosphere. Finally, in the summer of 2000, ESA launched the quartet of identical spacecraft known as Cluster II, four years after the original four spacecraft blew up during a failed rocket launch in June 1996. Satellites nicknamed Tango, Salsa, Rumba, and Samba now fly in a lopsided pyramid (or tetrahedron) formation, skimming the flanks of the magnetosphere in order to make simultaneous multipoint measurements and to create a three-dimensional picture of space weather activity.

In addition to these flagships, ISTP received significant contributions from other spacecraft, theory centers, and ground-based facilities run by Russia's Space Research Institute, NOAA-SEC, Los Alamos National Laboratory, Germany's Max Planck Institute, the U.S. Air Force, the Canadian Space Agency, the British

Antarctic Survey, and the National Science Foundation. In all, nearly a thousand scientists from more than 30 countries contributed observations to and analyzed data from the mission.

"ISTP has been one of the first programs to integrate experimental observations and theoretical physics successfully," notes Mario Acuña, project scientist for the ISTP program, based at NASA's Goddard Space Flight Center. "All of the ISTP elements were integrated into the ground system, which made the whole enchilada work as advertised. And we have proven to the world that theory is necessary to understand and extend observations where no spacecraft will ever go."

Dan Baker, one of the most vocal advocates of the coordinated studies of ISTP, notes how scientists can now study solar-terrestrial physics in microcosm and macrocosm. "Individually, the spacecraft contributing to ISTP act as microscopes, studying the fine detail of the Sun, the solar wind, and the boundaries and internal workings of Earth's magnetic shell," says Baker, head of the Laboratory for Atmospheric and Space Physics at the University of Colorado. "When linked with each other and the resources on the ground, they act as a wide-field telescope that sees the entire Sun-Earth environment."

The first real test of the ISTP philosophy came in January 1997, when the program's science team gathered at NASA's Goddard Space Flight Center in Maryland for a workshop. Assembled from laboratories around the world, the ISTP team was meeting to share past observations of coronal mass ejections (CMEs), flares, and bursts of high-speed solar wind and to coordinate future ones. It was a purposeful attempt to start evolving the field from uncoordinated, single-point studies of the space around the Earth to a global, system-wide perspective on how the Sun and Earth work together. But those few hundred scientists were not necessarily planning to do science in real time. During one of the science presentation sessions, solar physicist Don Michels of the U.S. Naval Research Laboratory walked to the front of the room and displayed images from earlier that day. He showed a time sequence where a cloud of plasma was bursting out from the Sun and head-ing toward Earth. While it did not appear to be an extraordinary CME, it was the first time the whole science team would watch such a storm develop in real time as it moved from Sun to Earth. Not only were they going to observe a space weather event in concert, they were going to do it in the same room.

When the CME arrived at Earth on January 10, the magnetic cloud of solar plasma took most of a day to pass. NASA scientist Keith Ogilvie estimated that the cloud was perhaps 30 million miles across, about the distance from Earth to Venus. And as his NASA colleague Robert Hoffman noted, the cloud "packed a one-two punch." Initially, the CME poured energy into the magnetosphere, initiating a magnetic storm and pumping up the energy in the radiation belts. Then, early on January 11, an unusually dense region of the CME cloud smacked the magnetosphere with a huge pressure pulse, with as many as 200 times more energetic particles packed into each cubic inch than in the rest of the cloud. "It was like Earth's magnetosphere was hit with a hammer," Hoffman said. "It rang the magnetosphere of Earth like a bell." It also appeared to shake up AT&T's Telstar 401 satellite, which suffered a catastrophic failure on January 11 (see Chapter 8).

The January 1997 storm made a big impression on the public as well as the scientific community. The loss of Telstar 401 and the fact that scientists had observed the event "from cradle to grave" made the event newsworthy. There were front-page stories in *The Washington Post, The Times of London,* and *The New York Times,* as well as broadcast reports on CNN, CBS, and National Public Radio. The media and the public were fascinated by the idea that the Sun may have killed a satellite. Just three months later, in April 1997, another CME was detected on its way toward Earth. Aided by some eager scientists still buzzing from the January storm, the news media announced that stormy times in space were imminent. This time there was some measure of panic in the public response. Nervous Californians called NASA to ask if the CME would trigger earthquakes. Rumors started to circulate on the Internet that the CME was actually a weapon that had been fired at the Earth by a UFO trailing comet Hale-Bopp. The eventual physical effects of the storm were minimal—auroras were visible in Boston and other northern U.S. cities, but no satellites or power grids had major failures. And, of course, no UFO emerged from behind the comet. But the April 1997 CME did have a practical effect. The UFO and earthquake incidents reminded scientists that there was a need for greater public understanding of the issue and the actual hazards associated with space weather. It also reminded those scientists that storms from the Sun are still difficult to predict.

The January 1997 event was proof of the ISTP concept. Images and observations from solar telescopes had been coupled with solar wind measurements; the information about the solar wind was compared with the effects measured in Earth's auroral zones, radiation belts, and magnetic tail. ISTP investigators had made the first complete, real-time study of a space weather event at a pace that Richard Carrington and Elias Loomis could have only dreamed about when they observed the great flare and aurora of 1859. And in a modern, Internet-wired world, they were able to share their observations and predictions—with colleagues around the world and with the taxpaying public—as the events were happening in real time.

"The international nature of the mission is a big plus and a big part of the experience," adds Pål Brekke, the European Space Agency's deputy project scientist for SOHO. The coordination of financial resources has allowed the spacecraft designers to think bigger; the coordination of scientific minds has allowed scientists to think more broadly. "A whole generation of American and European physicists are getting to know one another much better than they once would have only through meetings. It's a binding experience to work on a labor-intensive mission together."

As a result of a decade of cooperation and collaboration among scientists and nations, the paradigm of one-spacecraft/one-region studies has given way to global views of solar-terrestrial science. Some of NASA's newest Sun-Earth spacecraft—such as ACE, TRACE, the Fast Auroral Snapshot Explorer (FAST), and the Imager for Magnetopause to Aurora Global Exploration (IMAGE)—

are not formal members of ISTP, but each mission is collaborating with the existing ISTP missions because the scientific community now demands coordinated science.

In 2001 two new missions were launched and two others were being readied to both widen and focus that global view. On July 23, NOAA launched a new Solar X-ray Imager (SXI) as an instrument on the agency's newest GOES weather satellite, allowing solar physicists to collect minute-by-minute images of the Sun in X rays. Two weeks later the Jet Propulsion Laboratory and NASA launched the Genesis mission to capture samples of solar wind and stardust for three years before returning them to Earth. The Thermosphere-Ionosphere-Mesosphere Energetics and Dynamics (TIMED) mission of the Johns Hopkins University Applied Physics Laboratory and NASA was launched in December 2001 to study the uppermost reaches of Earth's atmosphere, where space weather phenomena may be coupled to the processes that drive our weather and climate. And finally, the High-Energy Solar Spectroscopic Imager (HESSI) was launched by NASA in February 2002 on a mission to explore the basic physics of particle acceleration and energy release in solar flares.

"Never before have scientists had such a complete set of tools with which to study the climax of a solar cycle," notes Dan Baker. "And never before have they had tools of such power and precision to study our most important star—the Sun—and our most important planet—Earth." By using these coordinated measurements—in the manner of atmospheric scientists who study global change through thousands of weather stations—space scientists are now dreaming about a day when they can make "weather maps" of approaching storms from the Sun. They are also vastly improving the models of space weather disturbances from solar outburst to impact on Earth's atmosphere.

The laws that govern electricity and magnetism are deceptively simple. There are only four of them, and they are named the

"Maxwell Equations" after the great nineteenth-century physicist James Clerk Maxwell. When Maxwell's work is combined with the equations of fluid flow for an electrically conducting gas, one arrives at the equations of MagnetoHydroDynamics, or MHD. In the minds of some scientists, MHD stands for "My Horrible Dream" because the resulting equations are so complex that they can be solved for only a very small number of special situations. Yet they are the equations that must be solved with ever more precision if scientists are going to understand and then predict the movement of energy and plasma from Sun to Earth.

Enter the computer. Any system of equations can be "solved" numerically by crunching the numbers to calculate the implications of the equations for a particular situation—that is, to create a simulation. The key is to have enough computing power to solve the equations within the desired time- and spatial scales. For years, MHD simulations of the magnetosphere were rather crude, idealized, and not practical or useful for advancing space weather prediction because few people had the time or the heavy-duty computers to handle all of the variables and permutations. In the mid-1990s that began to change. Computing power had advanced to the point where lots of realistic physics (though not all) could be built into the models. Models could suddenly be run in "real time" so that it would take a supercomputer one hour to simulate one hour of time. So various groups began to experiment with using real solar wind observations from spacecraft like Wind and ACE to drive their models, instead of invented solar wind conditions that are computationally easy to handle.

Such models have changed the way physicists see the invisible realm of the magnetosphere. "We have created the first global pictures of what is going on in the magnetosphere," notes Charles Goodrich of the University of Maryland, who has developed visuals of one of those models. Using a Cray C-90 and other powerful computers, Goodrich and colleagues (including one of the authors of this book) developed a series of scientific visualizations and a detailed analysis of the simulation results that depict for the first time how the magnetosphere responds to the shock of a real CME.

"Since we only have a few spacecraft, and they can only make point measurements," Goodrich adds, "this is the only way to look at the whole system."

As these new, more sophisticated simulations came onto the scene, scientists began comparing the results to real events. What they found was that the simulations work quite well at a global level. This is true even though the latest simulations still do not contain all of the physical processes that many space physicists consider important. Why do the simulations replicate reality as well as they do? The best answer could be that the solar wind is the key driver to magnetospheric activity, in contrast to terrestrial weather that is not so heavily forced on short timescales by solar activity. This also would explain why models of the magnetosphere based on chaos theory (which do not need to contain any plasma physics at all) reproduce a lot of the behavior of the magnetosphere using the solar wind input as a driver for the model.

This new understanding of the overwhelming role of the solar wind as a driver of the system, along with advances in our understanding of the basic physics of magnetic reconnection (the Cheshire cat of energy release in space plasmas), has led scientists to a grand idea. Shouldn't we simulate the entire system, from the surface of the Sun to the upper atmosphere of the Earth? Scientists are working on this challenge, trying to link together the flow of energy and plasma from one part of the system to the next. It will take a decade at least to accomplish the goal, but initial results have been quite promising, mainly because scientists now have coordinated coherent data from each part of the system.

"Most of the models already in use today do a reasonable job of predicting average conditions, but few of them take into account the dynamics and how quickly the system can change," says Terry Onsager, a researcher at the Space Environment Center who works to turn basic research findings into tools that forecasters can use. "But with the new stream of real-time measurements, we are beginning to synthesize mature models in order to give industry and the government the information it needs to work in space. Some of the new models that we are developing will allow us to visualize the radiation environment over vast regions of space and then specify and predict the conditions at any location. And they can take away some of the subjectivity of forecasting."

By the middle of the year 2000, the maturity and immaturity of the field of space weather prediction were on full display. Starting on July 11, 2000, forecasters at the Space Environment Center and scientists at ISTP mission control at NASA Goddard began closely watching solar active region 9077, a large group of sunspots with a complicated, gnarled structure. For four days they tracked a series of large flares and coronal mass ejections erupting from the region as it slowly lined up in the center of the solar disk for a head-on shot at Earth. Then, at 1024 Universal Time (6:24 a.m. U.S. Eastern Daylight Time) on July 14 (Bastille Day), the Sun unleashed a monstrous, long-lasting solar flare and with it a potent swarm of solar particles. A full "halo" coronal mass ejection was shot out from the Sun almost simultaneously, escaping the Sun at nearly 1,800 kilometers per second (4 million miles per hour).

The X rays and other light from the blast reached the cameras on SOHO and the flare sensors on the GOES weather satellites about 8 minutes later. Within 15 minutes after the flare, a stream of solar particles began bombarding the Earth's magnetosphere (moving at half the speed of light). Just 36 minutes after the flare, the ACE spacecraft was bombarded with so many particles that it lost its ability to track solar wind density and velocity. Spacecraft cameras trained on the Sun—from SOHO, TRACE, and Yohkoh—and on the auroral zones of Earth—from Polar and IMAGE—were blinded by the swarm of solar particles. One of the radio transmitters on the Wind spacecraft permanently lost about a quarter of its power, forcing controllers to switch to a backup system. After detecting the initial blast from the Sun, the very satellites that were supposed to monitor the incoming space weather were temporarily blinded by it.

About 26 hours after the Bastille Day flare, three shock waves arrived at Earth (they arrived twice as fast as the typical CME cloud). The shocks and the trailing blob of CME plasma smacked into the Earth's magnetosphere and compressed inside the geosynchronous orbit of many satellites. Because of the speed of the CME, forecasters had just 21 minutes between the detection of the storm by plasma and magnetic field detectors on the ACE spacecraft and the arrival at Earth; forecasters usually have 45 minutes to an hour of warning. By late afternoon on July 15, the largest magnetic storm since March 1989 began distorting the atmosphere around Earth, with a storm that reached the top of the scale on the index of space weather intensity (a 9 on the Kp scale of 1 to 9). Auroras raged in the skies over Europe, though few observers in the Americas saw the event at its peak because it occurred at 8 p.m. Eastern Time while the summer Sun was still marching toward the horizon. Had the storm occurred just a few hours later, observers would have seen auroras as far south as Florida and the Gulf of Mexico (see Figure 17).

Because space weather forecasters saw the storm brewing days ahead of time and issued alerts and warnings to let industry and the public prepare for the onslaught, damage from the Bastille Day storm of 2000 was mitigated. Electric currents caused by the magnetic storm caused voltage swings, tripping of capacitor banks, and damage to power transformers at more than a dozen electric power plants and companies in North America, but there were no blackouts. Had astronauts been working on the space station or the shuttle, they would likely have received the equivalent of a year's worth of allowable radiation in a matter of days. But no one was living on *Mir* or *Alpha*, and the shuttle was parked in Florida. Numerous commercial, military, and civilian science satellites suffered some damage to their solar panels or errors in their computer systems, but most equipment survived with some help from operators on the ground.

But the operators of Japan's Advanced Satellite for Cosmology and Astrophysics (ASCA) were not so fortunate. While the satellite did not suffer any major failures from the swarm of energetic

FIGURE 17. The Visible Imaging System on NASA's Polar spacecraft captured this image of the auroral oval at the peak of the July 2000 magnetic storm, around 8 p.m. U.S. Eastern Time on July 16. Auroras would have been visible across the entire continental United States had the storm occurred after dark. Courtesy of Visible Imaging System/University of Iowa and NASA.

particles, it was destroyed in a backhanded way. The energy from the flare and the CME increased the density of Earth's upper atmosphere so much that atmospheric drag began to pull the satellite down. The friction was so intense that it overwhelmed the momentum wheels that should have helped ASCA orient itself in space. Instead, the satellite went into a spin and lost power because its solar panels were no longer pointed toward the Sun. An astronomy satellite that studied black holes and distant galaxies was wiped out by the one star that it did not watch.

"This is a unique solar maximum, the most exciting in history," asserts George Withbroe, science director for NASA's Sun-Earth Connection program. "We have the most powerful fleet of spacecraft ever launched to study the Sun and its effects on Earth." Where once space scientists could only view the Sun for limited portions of the day from Earth or from orbits that descended into the planet's shadow several times a day, for the past five years they have been able to watch the Sun constantly. In a field where magnetic storms were typically detected after magnetometers on the ground started twitching, scientists have been able in recent years to use radars and spacecraft to pick up the signatures of those storms before they churn up the atmosphere. "The images and data are beyond the wildest expectations of the astronomers of a generation ago," Withbroe adds. And with the development of the Internet and the foresight of an open-data policy whereby all of the science teams share the observations from their instruments, just about all of the images and data are available for the world to see on the World Wide Web.

The ISTP era and the advent of 24-hour-a-day images of the Sun from SOHO have energized the field of solar-terrestrial science and provided scientists with a host of new discoveries and direct confirmations of long-held theories. In the past five years, space scientists have learned that the Sun squeezes a little bit of Earth's atmosphere into space with every CME blast. They have found that the atmosphere of the Sun occasionally configures itself in peculiar "S-shaped" structures before it launches some of its CMEs. Using sound waves from inside the Sun and beams of ultraviolet light propagating through the solar wind, researchers have begun to create images and models of the far side of the Sun, providing hope of space weather predictions made weeks in advance, rather than days. Combining auroral imagery with particle and magnetic field measurements in the cusps and tail of Earth's magnetosphere, space physicists have been able to gather firsthand evidence that magnetic reconnection is not just a blackboard theory and laboratory curiosity but an actual mechanism for moving energy in and around the Sun-Earth system. Using radio bursts and solar wind models, researchers are starting to make better predictions of the arrival of CMEs at Earth, narrowing the best estimates from a matter of days to hours. They have even been able to create coarse images of the invisible plasma hovering and swirling in the space around the Earth in the radiation belts.

"During the last solar max, we did not have the tools to follow solar activity in all layers of the solar atmosphere," says Pål Brekke of SOHO. "We now have all the data online via [the] Internet, so it's very easy to keep an eye on not only the Sun but also the effects on the Earth's environment."

The challenge of the next decade will be to turn those research findings into useful models and tools that can help the forecasters at SEC. "Within the research community, there has been continuous progress in studying and modeling the space environment," says Terry Onsager. "But very little of that research has made it into the space weather operations community. It's not so much that we are behind the times. It's just that clean, proven operations methods to handle all the new data are not available. Our job is to take the pulse of the research community and see where the real advances are made. Then we have to decide what we can move into operations."

In other words, not every discovery or advance in understanding necessarily translates into a tool that can be used, reliably and cost effectively, in day-to-day space weather forecasting. The single biggest problem is that most of the discoveries of the past decade have been made with scientific research satellites that were designed for discovery and exploration, not for steady, daily monitoring of conditions. Every time researchers discover the "next big advance" in space weather forecasting, they are usually just offering a proof that a certain type of observation with a certain satellite in a certain situation can work. They make a discovery, suggest that it might be useful, and then move on to the next research interest.

By the very nature of the way science is funded, scientific advances and novel concepts that could turn into forecasting tools are often filed away in the data bins before they can be turned into reliable, reproducible applications for SEC. It is a troublesome quirk of federal science programs that the government will pay billions of dollars for "basic" research projects in space sciences, but the practical applications are often left for industry, which doesn't always answer the call. It certainly isn't in the financial interests of a satellite company to spend its money to monitor the space environment and share that information for free with competitors. And government research programs rarely promote or financially reward scientists for the application of basic research, at least not in ways that draw many scientific minds to work on the "applied" side. For the scientific researchers trying to maintain a lab, a staff, and a standard of living, it is more compelling and self-preserving to propose "new" science exploration to the government budget managers than to admit that the old questions still need answers, that the old data and observations have not yet ripened into beneficial applications. NASA and the National Science Foundation, by their very charters, are not supposed to get involved with "monitoring" projects. So monitoring the weather—on Earth and in space—becomes a dirty word, an activity that is shunned because no one wants to be underfunded or taken for granted.

So as a practical matter, space weather forecasting is currently tied to the uncertain life of research satellites and instruments that may be gone before they can be fully integrated into forecasting schemes. For instance, while the SOHO and ACE missions have revolutionized the world's view of the Sun and solar wind, the SEC forecasting team must be prepared to live without those tools. Why? Research missions are funded for limited periods of time. Once the major research objective is reached, the agencies are looking to the next big advance in science. Missions that SEC counts on—but cannot control or afford—are subject to frequent budget reviews and constant threats of cancellation so that the research agencies can tread on the cutting edge. But forecasting does not require constant advances in technology; it requires tried and tested and reliable observations and models. Until NOAA-SEC knows it can fly a spacecraft that will make the same measurements as SOHO, the staff will have to prepare itself for life without one of their most useful tools.[1]

Despite the tensions of basic research versus applied science, despite the growing pains of turning discoveries in individual regions of space into a new understanding of the Sun and Earth as a system, researchers remain transfixed by their star. Every new answer about how the Sun or magnetosphere works leads scientists to a new set of questions. It is like peeling the proverbial onion: peeling any layer away reveals another one just as complex and compelling. "Simply watching the never-ending procession of changes on the Sun in full-resolution movies still takes my breath away, even after five years of it," says Joe Gurman, NASA's project scientist for the SOHO mission. "Every time we image the Sun on any new timescale, we discover that things change on those timescales and on all longer ones that we can visualize."

Epilogue 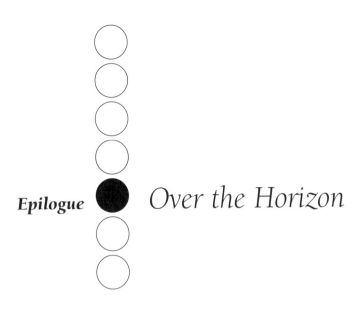 *Over the Horizon*

It's hard to make predictions—especially about the future.
                                     Attributed to Yogi Berra

The year is 2012, the peak of solar maximum 24. Most citizens of the industrialized nations carry personal communicators, hybrid devices that serve as a cell phone with worldwide access, a pager, an instant messenger, and a Web browser with high-speed access to a satellite-based Internet. Any time of day, from anywhere on the planet, the modern citizen can check her bank account, talk to her son about his science fair project, order her groceries, and submit her plan for restructuring the sales division of her company. She is never out of touch because the personal communicator routes its signals through a string of geosynchronous satellites hovering around the world. She takes for granted the idea that she can work or chat with distant colleagues all the time, whether on the beach, in the office, or on the front porch of Grandma's 200-year-old farmhouse in the sticks.

Our twenty-first-century citizen knows that when her plane lands tonight—she is cruising on a supersonic jet from Beijing to New York along the polar route—her home will be coolly air conditioned, thanks to the solar panels on her rooftop. Her neighbor won't be so lucky, as his muggy house is still tapped into the local power grid. Since no electric power plants have been built in the past 10 years, he must endure rolling blackouts at least once a week. Lightning storms and the occasional geomagnetic storms play havoc with an overloaded system.

Pulling up the day's news on her communicator, our modern citizen notices that plans for the international Moon base are proceeding rapidly, with move in scheduled for 2017. But she also reads that NASA and the European Space Agency are struggling with the design of the spacecraft that will carry the first human crew to Mars, as major questions remain about the effects of long-term human exposure to radiation in space. In a small side article, she notes that NASA's Voyager 2 spacecraft has passed the heliopause, the outermost edge of our solar system. She opens her e-mail account to see an offer to participate in an initial public stock offering for a Russian space tourism company that is planning to launch an orbiting hotel for civilian passengers; they also plan to offer solar sailing cruises around the Moon. The proposed hotel will include private rooms for couples, with a three-day visit to low-Earth orbit costing about $500,000 per person. She saves the message, thinking about how she might surprise her husband for his birthday.

At about the time she summons the terrestrial weather report, the plane's pilot pipes in over the intercom. As predicted by the Space Environment Center, the Sun has erupted with activity this afternoon. The flare is near the top of the scales of intensity. A solar particle storm should begin in about 10 to 15 minutes, so the plane will have to divert from its polar route. The disturbed ionosphere will not allow high-frequency radio communications and airlines require continuous contact with aircraft. And ever since that big class-action lawsuit in 2009, the airlines have been scrupulous in avoiding any chance of exposing passengers and crew to unhealthy doses of radiation. The plane is diverted to Seattle, and the flight will now land at least four hours later than planned.

The coronal mass ejection (CME) that ripped off the Sun in conjunction with the flare is traveling at about 5 million miles per hour, with arrival expected in 19 or 20 hours—in the middle of the next workday on the East Coast of the United States. Forecasters are warning the power companies to brace for a wicked magnetic storm, which likely means that many cities along the East Coast will be forced to endure preventative rolling blackouts, disrupting commerce for the day. The NASDAQ has already declared a day off from stock trading tomorrow, as there is too much risk of garbled transactions now that all of the "buy" and "sell" orders are relayed by satellite. Other markets are beginning to follow suit.

About an hour later, the captain adds more bad news to our modern citizen's day. He tells passengers that the solar particle event has temporarily crippled the radio communications system at the airport and that Global Positioning System signals are distorted, so air traffic controllers are having difficulty managing the flow of aircraft into a foggy airport in Seattle. Traffic is backing up, and they may have to divert the plane to Portland or Vancouver. It looks like our modern citizen will be spending another night on the road and will miss her daughter's softball game tomorrow.

Our daydream for the year 2012 is part cautionary tale, part crystal-ball speculation—though not quite the wondrous leaps made by Jules Verne or Arthur C. Clarke. The way we use technology to work and play is evolving daily, and while the specifics of a device like a personal communicator are yet to be worked out, it is easy to speculate that we will be communicating in some unthinkable way just a decade from now. After all, just a decade ago the World Wide Web barely existed (at least from a public point of view) and cellular phones operated more like children's walkie-talkies. Today, a Finn in Helsinki can buy a Coke from a vending machine by dialing a number on his cell phone and the cost of the Coke appears on his phone bill.

Our scenario for 2012 is not such a stretch. Every one of the space and communications phenomena described in our future citizen's life has already been discussed in laboratories or industry boardrooms. We know that many industries are interested in moving messages and data through a satellite-based infrastructure—that is, routing the information through radio beams to satellites rather than expensive wires on the ground. And in order to build that infrastructure in a cost-efficient way, satellite companies will have to make choices about the hardiness of their spacecraft. The current trend in the industry is to build and launch more satellites with less protection, to expect some catastrophic failures, and to replace the losses with other satellites in waiting. The next solar maximum will certainly put that business plan to the test.

As for electric power systems, all one needs to do is pick up a daily newspaper to see that a new energy crunch is developing in North America. With growing interest in environmental issues over the past three decades and the explosion of the suburban "not-in-my-backyard" mindset, very few new power plants have been built. The public and the power industry have been too afraid of nuclear accidents to build new fission plants, too concerned about clean air (or, conversely, corporate profits) to build new fossil fuel-powered plants, and too impatient to work out the cost issues for solar, geothermal, and wind power. Yet demand for electricity has increased exponentially, and the supply has remained stagnant. The existing electric power systems of the world are already being stressed by surging demand, and when the system is stressed, it is more vulnerable to equipment failures and blackouts due to terrestrial and space weather.

With the demise of the Soviet Union, the polar airline routes between North America and the booming economies of East Asia have become busy highways, cutting the flight time significantly. But these highways are vulnerable to space weather, and unexpected diversions to Tokyo and other landfalls are a fact of life. Those passengers who are inconvenienced today probably have no idea that a warning from the Space Environment Center was the real reason for the diversion.

Space tourism may not boom in time for the twenty-fourth solar maximum, but the time is rapidly approaching when the everyday citizen might pay to venture into the realm of the astronauts. Dennis Tito opened the door by buying a trip on International Space Station *Alpha* in the spring of 2001. In Russia, emerging capitalism and a surplus of unused space expertise will surely lead to new space ventures. Even in the United States, the U.S.-based Planetary Society and other partners had scheduled for 2001 the launch of the first solar sail, a technology that uses the pressure of sunlight—rather than rockets and liquid fuel—to push and accelerate spacecraft just as the wind pushes a sailboat across the sea. The National Aeronautics and Space Administration (NASA), the National Oceanic and Atmospheric Administration (NOAA), and other federal agencies were also making plans for incorporating solar sails into future missions, perhaps for the next fleet of space weather monitoring craft. Solar system cruising is on the horizon.

In order to prepare for this future, NASA, NOAA, the European Space Agency, the National Science Foundation, and the U.S. Air Force are queuing the next generation of spacecraft to study storms from the Sun. In an ambitious plan that has been percolating since 1999, NASA and its partners have been designing and proposing a program to both study and monitor the Sun-Earth system for practical purposes. They call the program "Living with a Star." Instead of starting from purely scientific interests, scientists and engineers have been considering the utilitarian issues, the impacts and effects of space weather. Basic researchers are working with forecasters, military and commercial satellite specialists, and other industry and applied science groups to define the data sets and the spacecraft that can best improve our understanding of space weather.

Early plans call for five different types of spacecraft. One satellite will observe the Sun and its interior workings from much closer range than ever before (about one-third the distance from the Earth to the Sun). A pair of "sentinels" will fly at two wide angles ahead and behind the Sun-Earth line, allowing scientists to see CMEs and other phenomena developing in three dimensions. A solar wind monitor—comparable to ACE—will fly on a solar sail. A pair of satellites will crisscross Earth's radiation belts to study the development and movement of satellite-damaging particles. And a constellation of miniature satellites will swirl in low-Earth orbit, studying how space weather connects to the Earth's ionosphere and atmosphere. Launches are expected to begin late in the first decade of the twenty-first century, and mission managers hope to have the whole fleet in place to watch the twenty-fourth solar maximum, much as the International Solar-Terrestrial Physics program has allowed researchers to dissect the current solar maximum.

Computer simulations of unprecedented sophistication will complement these new data. Currently, several groups of physicists are working to produce simulations that will allow, for example, solar observations to predict the density, speed, and magnetic field inside a CME as it slams into the magnetosphere. By 2012, space weather will likely be more like current-day meteorology, with more spacecraft and ground stations available to monitor all the key parts of the system and end-to-end models of the flow of energy from the solar interior to the Earth's surface. Space weather forecasting—assuming that governments continue to fund the satellites and observatories—should improve in its accuracy as scientists come to better know the conditions that provoke storms on the Sun and in the space around the Earth.

But many questions need to be answered before we can get truly comfortable living in the atmosphere of our nearest star. "Is a CME just a cute trick, or is it something the Sun has to do?" asks JoAnn Joselyn, a longtime scientist at the Space Environment Center. "And how do solar events evolve? Is there steady progress, or is it all a big burst?" Researchers also are perplexed by the day-to-day workings of the Sun, such as what makes the corona so much hotter than the visible surface and what accelerates the solar wind. Closer to Earth, space physicists wonder what parts of a

CME really affect the magnetosphere and how the magnetosphere and radiation belts can focus pedestrian plasmas and disorganized solar energy into raging, satellite-killing particles.

"In addition to understanding the relationship between flares, CMEs, and proton events—we still don't really know what causes them, I think—we also need to understand what is causing the solar activity to vary on decadal and century timescales," says Päl Brekke. "The key to these questions lies beneath the surface of the Sun."

The biggest mysteries of all rest on the borders between solar physics and climatology and between space physics and medicine. The atmosphere of Earth is warming for reasons we have not yet been able to fully decipher. Human activity certainly plays some role in climate change, as do other processes of wind and water. But since the Sun is the primary driver of climate on Earth, doesn't it stand to reason that solar variability must somehow affect the habitability of the planet? Only time and observation can tell us.

As for the biological aspect of space weather, we are forced to consider—as we become a space-faring race—whether we can survive out there. Science fiction conjures up all manner of dangerous aliens and starbursts and quirky, mind-bending ripples in space and time. But the more immediate danger comes from our own Sun. Humans and other earthly life forms evolved within the protective shell of the magnetosphere and the atmosphere, shielded from most of the harmful solar and cosmic radiation. As we prepare to venture beyond the life-preserving cocoon, we will quickly find out how much our life depends on the invisible magnetic force field around our planet. Perhaps we really do need shelter from the storms from the Sun.

Appendix A Selected Reading

Appell, D. "Fire in the Sky." *New Scientist*, February 27, 1999.

Baker, D. N., J. H. Allen, S. G. Kanekal, and G. D. Reeves. "Disturbed Space Environment May Have Been Related to Pager Satellite Failure." *Eos, Transactions, American Geophysical Union*, Vol. 79, No. 40, October 6, 1998.

Barish, R. J. *The Invisible Passenger: Radiation Risks for People Who Fly*. Madison, WI: Advanced Medical Publishing, 1996.

Biermann, L. F., and R. Lust. "The Tails of Comets." *Scientific American*, Vol. 199, No. 4, October 1958.

Bone, N. *The Aurora: Sun-Earth Interactions*. New York: John Wiley & Sons, 1996.

Brekke, A., and A. Egeland. *The Northern Light: From Mythology to Space Research.* New York: Springer-Verlag, 1983.

Brooks, J. "Reporter at Large: The Subtle Storm." *The New Yorker,* February 27, 1959.

Burch, J. L. "The Fury of Space Storms." *Scientific American,* Vol. 264, No. 10, April 2001.

Calder, N. *The Manic Sun: Weather Theories Confounded.* London: Pilkington Press, 1997.

Calvin, W. H. *How the Shaman Stole the Moon: In Search of Ancient Prophet-Scientists from Stonehenge to the Grand Canyon.* New York: Bantam Books, 1991.

Clarke, A. C. *The Wind from the Sun: Stories of the Space Age.* New York: Harcourt Brace Jovanovich, 1972.

Dwivedi, B. N., and K. J. H. Phillips. "The Paradox of the Sun's Hot Corona." *Scientific American,* Vol. 264, No. 12, June 2001.

Eather, R. H. *Majestic Lights: The Aurora in Science, History, and the Arts.* Washington, D.C.: American Geophysical Union, 1980.

Eddy, J. A. "Probing the Mysteries of the Medicine Wheels." *National Geographic,* January 1977.

Eddy, J. A. "The Case of the Missing Sunspots." *Scientific American,* Vol. 236, No. 5, May 1977.

Fordahl, M. "32 Million Pagers Go Silent." Associated Press, May 20, 1998.

Golub, L., and J. M. Pasachoff. *Nearest Star: The Surprising Science of Our Sun.* Cambridge, Mass.: Harvard University Press, 2001.

Haines-Stiles, G. H., and E. Akuginow. *Live from the Sun Fact Book.* Morristown, NJ: Passport to Knowledge, 1999.

Harford, J. J. "Korolev's Triple Play: Sputniks 1, 2, and 3" from *Korolev: How One Man Masterminded the Soviet Drive to Beat America to the Moon.* New York: John Wiley & Sons, 1997.

Hufbauer, K. *Exploring the Sun: Solar Science Since Galileo.* Baltimore: Johns Hopkins University Press, 1991.

Jago, L. *The Northern Lights: The True Story of the Man Who Unlocked the Secrets of the Aurora Borealis.* New York: Alfred A. Knopf, 2001.

Kippenhahn, R. *Discovering the Secrets of the Sun.* New York: John Wiley & Sons, 1994.

Lang, K. R. *Sun, Earth and Sky.* New York: Springer-Verlag, 1997.

Lawn, V. "Aurora Borealis Blacks Out Radio." *The New York Times*, February 11, 1958.

Littman, M. K., K. Willcox, and F. Espenak. *Totality: Eclipses of the Sun.* New York: Oxford University Press, 1999.

Loomis, E. "The Great Auroral Exhibition of August 28th to September 4th, 1859." *American Journal of Science and Arts*, Vol. 78, No. 82, 1859.

Maran, S. P. *Astronomy for Dummies.* New York: IDG Books Worldwide, 2000.

McIntosh, P. S. "Did Sunspot Maximum Occur in 1989?" *Sky & Telescope*, January 1991.

Monastersky, R. "The Sunny Side of Weather." *Science News*, Vol. 146, December 3, 1994.

The New York Times. "*Aurora Borealis* Gives City a Show as Sun Spots Disorganize Radio." September 19, 1941.

The New York Times. "Northern Lights Display: Telegraph and Cable Lines Suffer by Electrical Disturbance." November 1, 1903.

The New York Times. "Sun-Spot Tornado Disrupts Cables, Phones and Telegraph for 5 Hours." March 25, 1940.

Odenwald, S. F. *The 23rd Cycle: Learning to Live with a Stormy Star.* New York: Columbia University Press, 2001.

Parker, E. N. "The Physics of the Sun and the Gateway to the Stars." *Physics Today*, June 2000.

Roberts, W. O. "Corpuscles from the Sun." *Scientific American*, Vol. 192, No. 2, February 1955.

Rust, D. M. "The Great Solar Flares of August 1972." *Sky & Telescope*, October 1972.

Sawyer, K. "Earth Takes a 'One-Two Punch' from a Solar Magnetic Cloud." *The Washington Post*, January 23, 1997.

Schaefer, B. E. "Solar Eclipses that Changed the World." *Sky & Telescope*, May 1994.

Sky & Telescope. "February's Great Multicolored Aurora." April 1958.

Suess, S. T., and B. T. Tsurutani, Eds. *From the Sun: Auroras, Magnetic Storms, Solar Flares, Cosmic Rays*. Washington, D.C.: American Geophysical Union Books, 1998.

Sullivan, W. *Assault on the Unknown: The International Geophysical Year*. New York: McGraw-Hill, 1961.

Tribble, A. C. *The Space Environment: Implications for Spacecraft Design*. Princeton, NJ: Princeton University Press, 1995.

Van Allen, J. A. *Origins of Magnetospheric Physics*. Washington, D.C.: Smithsonian Institution Press, 1983.

Verschuur, G. "The Day the Sun Cut Loose." *Astronomy*, August 1989.

Appendix B Selected Web Sites

This book has a companion site on the World Wide Web where you can view images and movies related to space weather, read reviews of the book, and communicate with the authors. Please visit *http://www.stormsfromthesun.net*.

Below you will find a selection of web sites that we find to be among the most interesting and informative about space weather and its effects.

Overviews of the Sun, Space Physics, and Space Weather

Exploration of the Earth's Magnetosphere
http://www.phy6.org/Education/Intro.html

Interesting Facts and Educational Materials About Space Weather
http://www.ips.gov.au/papers/

Live from the Sun
http://www.passporttoknowledge.com/sun

Living in the Atmosphere of the Sun: The Space Weather Center
http://istp.gsfc.nasa.gov/exhibit/

Mission to Geospace
http://istp.gsfc.nasa.gov/istp/outreach/

SolarMax IMAX Film
http://www.solarmovie.com/

Space Weather at Windows to the Universe
http://www.windows.ucar.edu/spaceweather

Stanford Solar Center
http://solar-center.stanford.edu/

Yohkoh Public Outreach Project
http://www.lmsal.com/YPOP/index.html

Real-Time Space Weather Observations

The Space Weather Bureau
http://spaceweather.com/

Space Weather Now
http://www.sec.noaa.gov/SWN/

What's Happening Today in Space?
http://www.windows.ucar.edu/spaceweather/more_details.html

Major Space Weather Programs

European Space Agency
http://www.estec.esa.nl/wmwww/spweather/

High Altitude Observatory
http://www.hao.ucar.edu/

NASA Sun-Earth Connections Science Theme
http://sec.gsfc.nasa.gov/

National Solar Observatory
http://www.nso.noao.edu/index.html

NOAA Space Environment Center
http://www.sec.noaa.gov/

Scientific Committee on Solar-Terrestrial Physics (SCOSTEP)
http://www.ngdc.noaa.gov/stp/SCOSTEP/scostep.html

U.S. Air Force Battlespace Environment Division
http://www.vs.afrl.af.mil/Division/battle.htm

Other Interesting Space Sites

Astronomy Picture of the Day
http://antwrp.gsfc.nasa.gov/apod/

Aurora's Northern Lights
http://climate.gi.alaska.edu/curtis/curtis.html

Bad Astronomy
http://www.badastronomy.com/

Eclipse Home Page
http://sunearth.gsfc.nasa.gov/eclipse/eclipse.html

From Stargazers to Starships
http://www.phy6.org/Stargaze/Sintro.htm

The Galileo Project—Rice University
http://es.rice.edu/ES/humsoc/Galileo/

Mr. Eclipse
http://www.MrEclipse.com

In-Flight Radiation Protection Services
http://members.tripod.com/robbarish/

The Planetary Society—Solar Sail Project
http://www.planetary.org/solarsail/

Satellite News Digest—Satellite Outages and Failures
http://sat-nd.com/failures/index.html

Shooting the Aurora Borealis
http://www.ptialaska.net/~hutch/aurora.html

Appendix
C Acronyms and Abbreviations

AC	alternating (electrical) current
ACE	Advanced Composition Explorer satellite
ACRIM	Active Cavity Radiometer Irradiance Monitor
Alpha	unofficial name of the International Space Station
ASCA	Advanced Satellite for Cosmology and Astrophysics
AURA	Association of Universities for Research in Astronomy
BAS	British Antarctic Survey
BBC	British Broadcasting Company
C^{14}	carbon-14 isotope
CBC	Canadian Broadcast Company

CME	coronal mass ejection
DC	direct (electrical) current
DoD	Department of Defense (U.S.)
DOSMAP	dosimetric mapping experiment
DST	disturbance storm-time index
ERB	Earth Radiation Budget experiment
ESA	European Space Agency
EVA	extravehicular activity; also known as a space walk
FAA	Federal Aviation Administration (U.S.)
FUSE	Far Ultraviolet Spectroscopic Explorer
GIC	geomagnetically induced current
GOES	Geostationary Operational Environmental Satellites
GPS	Global Positioning System
HAO	High Altitude Observatory (U.S.)
HF	high frequency
IMF	interplanetary magnetic field
IGY	International Geophysical Year
ISS	International Space Station
ISTP	International Solar-Terrestrial Physics program
LASCO	Large-Angle Spectrometric Coronagraph
LEO	low-Earth orbit
LORAN	Long-Range Navigation system, maintained by the U.S. Coast Guard
LWS	Living with a Star program (NASA)
MHD	MagnetoHydroDynamics
MW	megawatts of electric power

NASA	National Aeronautics and Space Administration
NOAA	National Oceanic and Atmospheric Administration
NPR	National Public Radio
NRC	National Research Council (U.S.)
NRL	Naval Research Laboratory (U.S.)
NSF	National Science Foundation
rem	Roentgen equivalent man, a measure of radiation absorption by the human body
SAA	South Atlantic Anomaly
SAMPEX	Solar, Anomalous, and Magnetospheric Particle Explorer
SEC	Space Environment Center
SEU	single-event upset (in satellite computers)
SMM	Solar Maximum Mission
SOHO	Solar and Heliospheric Observatory
TRACE	NASA's Transition Region and Coronal Explorer
UARS	Upper Atmospheric Research Satellite
UFO	unidentified flying object
USAF	United States Air Force
USGS	U.S. Geological Survey
UT	Universal Time, also known as Greenwich Mean Time
UV	ultraviolet radiation

Endnotes

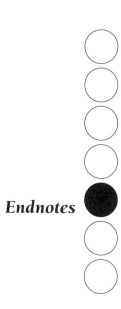

Chapter 1: The Day the Pagers Died

1. Available on the World Wide Web at *http://www.agu.org/sci_soc/articles/eisbaker.html*.
2. At the peak of the solar cycle in 2000, PanAmSat felt compelled to post a "statement on solar activity" on the front of its corporate web site. Acknowledging the threat of solar activity and solar maximum, the company noted: "As a global satellite operator, we are prepared to fly our satellites in the varied conditions of space. The recent solar flares, although powerful, are a highly predictable 'weather' condition, and we expect no impact to our fleet. . . . Based on our knowledge of the effects on spacecraft of environmental conditions in space, we do not believe these conditions will have any material adverse effect on our fleet. To our knowledge, no such effects were suffered by any communications satellite during the last peak period of solar flare activity. . . . PanAmSat's satellites are designed with shielding to protect against most of the major conditions that will be created during peak periods of solar flare activity."

Chapter 2: Sun-Eating Dragons, Hairy Stars, and Bridges to Heaven

1. It is an unlikely coincidence of nature that the Sun and the Moon are almost exactly the same size from the perspective of Earth. The Sun

is about 400 times larger than the Moon, but it is also 400 times farther away.
2. While plasma is extremely rare on Earth—you can only find it in lightning, fluorescent lights, and candle flames—it is actually the cosmic lifeblood. Scientists estimate that more than 95 percent of the visible mass of the universe is comprised of plasma.
3. To his credit, Celsius and protégé Olof Hiorter would later use the deflection of compass needles to show that auroras were related to magnetism.

Chapter 4: Connecting Sun to Earth

1. As noted in Dava Sobel's 1995 book, *Longitude*, Graham was a patron and key supporter of John Harrison, who developed the chronometer that solved the longitude problem of navigation.
2. Birkeland was a colorful character. Science was not well funded around the end of the nineteenth century, so Birkeland engaged in a variety of activities to raise money. One of his ideas was to build and sell an electromagnetic cannon. Unfortunately, during the first demonstration of the cannon to an audience of dignitaries and representatives of industry (including representatives of Krupps, the German weapons maker), it blew up when he closed the switch. Birkeland nonetheless used this failure to found a new industry. He teamed up with an engineer, Sam Eyde, whom he'd met a dinner party and who was looking for a way to create artificial lightening in order to generate nitrogen compounds. Their partnership led to the invention of the electromagnetic furnace for producing nitrogen fertilizers. Together, Birkeland and Eyde founded Norsk Hydro, which began commercial production of fertilizer, and both became wealthy men. Today, Birkeland is immortalized by his portrait on the Norwegian 200 Kroner note.
3. In the wake of the 1859 storm, Loomis began to construct the first of several maps of the frequency of the appearance of northern lights by geographic region. Scientific data gathering confirmed what the unscientific observer could have suspected—auroras were most frequent in the high latitudes of Scandinavia, Canada, Alaska, and Greenland, forming a ring that is tilted slightly toward North America due to the location of Earth's magnetic North Pole in northern Canada.

4. Not content with his laboratory experiments, Birkeland undertook several expeditions to northern Scandinavia to establish observatories for studying the aurora. Since one needs a dark sky to see the northern lights, Birkeland made his expeditions during the winter, which nearly cost him his life. He also wanted to spend the winter on a mountaintop in Finland because he had heard that the aurora would come down and touch the mountains. Alex Dessler pointed out that if the story was true, Birkeland would have been engulfed by the aurora, but Birkeland was not the least bit concerned.

Chapter 5: Living in the Atmosphere of a Star

1. As author and astronomer Steve Maran points out, the Sun is by no means average: "The vast majority of all stars are smaller, dimmer, cooler, and less massive than our Sun." So if you are measuring the Sun against the extremes of stars, it is somewhere in the middle. But if you measure it against the population of the universe, it is rather large.

Chapter 6: The Cosmic Wake-Up Call

1. Since scientists began making X-ray measurements of flares in 1976, only two flares have surpassed the potent blast of March 6, 1989. On August 16, 1989, the Sun popped off an X-20 flare, the highest classification. The most potent flare ever recorded sprang from the Sun on April 2, 2001, as described in the prologue of this book. Officially classified as an X-20 flare, the flare was actually off the scale. Of the 23 largest flares on record, only 4 occurred prior to the peak of the solar cycle, with the rest occurring at the maximum or on the way down from the height of activity. The worst year for large solar flares was 1991, with 8 of the 23 flares in the record books. The latest flare in a solar cycle occurred in May 1984, more than four years after solar maximum. The November 1997 flare was the earliest in a cycle to make the list—only 18 months after solar minimum.
2. There is not just one international standard index of magnetic storms. The 1989 storm is either the first, second, or third most potent magnetic storm since scientists began regular reporting in the 1930s, depending on which index you choose.

3. According to Boteler, the first reported effect on electric power systems occurred on March 14, 1940, though GICs probably affected power lines from the very first installation. On Easter Sunday 1940, a major magnetic storm sent stray currents through telegraph and power systems in North America. More than 20 power companies in New England, New York, Pennsylvania, Minnesota, Ontario, and Quebec reported disturbances.

Chapter 7: Fire in the Sky

1. Anticipating the solar maximum period predicted for 1957 to 1958, the International Council of Scientific Unions in 1952 proposed a comprehensive series of global geophysical activities modeled on the International Polar Years of 1882 to 1883 and 1932 to 1933. The International Geophysical Year (IGY), as it was called, was planned for the period from July 1957 to December 1958. The intention was to allow scientists from around the world to take part in a series of coordinated observations of various geophysical phenomena. Initially, scientists from 46 countries originally agreed to participate; by the close of the IGY, 67 countries had contributed. IGY activities spanned the globe from the North to the South Poles, with much of the work conducted in the arctic, antarctic, and equatorial regions.
2. The Explorer I satellite—the first launched by the United States—had been orbiting the Earth for 11 days when the February 1958 storm arrived. James Van Allen wrote an article about the event as recently as 1999.
3. In November 1903, *The New York Times* reported that all of the streetcars of Geneva, Switzerland, were "brought to a sudden standstill" for a half-hour by a potent magnetic storm. The event caused "consternation at the generating works, where all efforts to discover the cause were fruitless." According to British meteorologists of the time, the streetcar and telegraph troubles were attributable to sunspots and solar activity, "which would also account for the unusual wet season now being experienced."

Chapter 8: A Tough Place to Work

1. In 1999 the Department of Energy's Sandia National Laboratory announced that it was awarding a $1 million contract to a company to develop a radiation-hardened Pentium-powered computer board. The

board would be close to seven times as powerful as today's space-based computer systems, which typically lag years behind the power and capability of even commercially available personal computers.
2. The angst over sharing information about space weather and satellites is not limited to private or Western industry. In February 2001, Russia's FSB domestic security agency (the successor to the KGB) arrested a physicist in Siberia for selling space research information to China. Valentin Danilov, head of the Thermo-Physics Center of the Krasnoyarsk State Technical University, was charged with treason for trying to sell research on the effects of space weather on satellites. His research had been a state secret until 1992, but it was in the public domain for most of the past decade. Chinese colleagues had taken an interest in his work, perhaps too much interest for old Cold Warriors who still feel that satellite technology is a national interest worth protecting.

Chapter 9: Houston, We Could Have a Problem

1. Gautam Badhwar died suddenly in August 2001 at the age of 60. He was in the midst of leading two major experiments in radiation effects in space, with instruments on the International Space Station and the Mars Odyssey spacecraft. He will be greatly missed.
2. Rem stands for Roentgen equivalent man, a measure of the amount of radiation it takes to affect organic materials and human tissues. Radiation doses are measured in several different units. The amount of radiation per unit of mass is often measured in rads and Grays (100 rads equals 1 Gray); biological exposures are now typically measured in sieverts, with 1 sievert being equivalent to 100 rem. One rad of radiation typically equals between 2.2 and 2.5 sieverts, depending on the spectrum of the radioactive particles.
3. The Mars Odyssey spacecraft that traveled to the red planet in 2001 carries a set of instruments designed by Gautam Badhwar and colleagues to measure the cosmic and solar radiation that reaches Mars's atmosphere. During the 30 months of Odyssey's main mission, the Mars Radiation Environment Experiment (MARIE) is supposed to collect data on the amount and type of radiation deposited in the Martian atmosphere, for use in preparing for future human flights. However, when Odyssey arrived at Mars in October 2001, MARIE's software was malfunctioning. Mission operators and engineers were working to restore the instrument, which did not appear to be catastrophically damaged.

Chapter 10: Seasons of the Sun

1. The largest solar maximum in recorded history occurred in March 1958, during solar cycle 19, when the sunspot number reached 201. About four major solar flares occurred during that maximum. In solarmax 20 (1968) the sunspot number peaked at 110, with just one major flare, and in solarmax 21 (1979) the sunspot count reached 164, but there were no major flare events. In the now-famous solar cycle 22 (1989), the peak sunspot number reached 158, modest compared to 1958, but the number of large solar flares was 8.

Chapter 11: The Forecast

1. In the summer of 2001, at the recommendation of a panel of space and solar physicists, NASA announced the cancellation of the International Solar-Terrestrial Physics program. Despite the fact that all of the spacecraft were working, the agency decided that official coordination of the individual missions was a scientific luxury it could no longer afford in a tough budget environment. The Wind spacecraft was slated for a sort of space storage as a backup to ACE, and NASA withdrew its support for Wind's science investigators and for the American participants in Japan's Geotail mission. IMP-8, a venerable craft that had contributed data on interplanetary space for 25 years, was shut down, and the SAMPEX and FAST missions to study auroras and the inner magnetosphere were scheduled for shutdown in 2003 and 2004, respectively. SOHO, Polar, and Cluster were funded at reduced budget levels and encouraged to coordinate their work with other missions. But they were not given much funding to pay for the tasks of coordination, and funding for some of the key elements of the ISTP success story—the theory and modeling programs, the data centers, and the ground-based observatories—was almost entirely cut off.

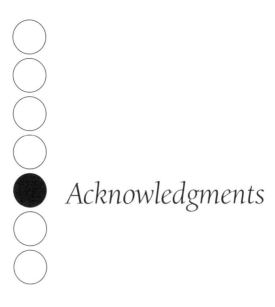

Acknowledgments

As we have been reminded many times, you never finish, you just run out of time. And so it is with this book. It took two years to plan and to convince ourselves that there was a story to be told, with a little push from colleague Pat Reiff of Rice University, who once suggested "we need to find a writer to put together a book on space weather." It took another two years of research, interviewing, outlining, thinking, writing, revising, and rewriting . . . with a few pauses for career moves and the births of two children. The process could continue endlessly, just as any scientific inquiry always leads to a new set of questions. But it is time to share our limited view of a burgeoning and exciting new science of our time.

It has been an odyssey for two novice authors but one we would both take again. Along the way we have found many great stories and interesting people, but it is the nature of storytelling that some things get in and some get left out. This book is intended as a mere introduction to an intriguing subject and not the definitive, comprehensive work that tells the whole truth. There are fine books and web sites already written about the inner workings of the Sun and a few very good ones about the aurora as well. But ours is an attempt to pull the Sun and Earth into the same field of view, to reveal our planet and our star as part of an interconnected system that brings a lot more to our life than just daylight and sunburns. Painting the big picture of space weather and its implications means that we had to use broad brush strokes, so we must apologize in advance to our colleagues for the discoveries, the great minds, and the scientific nuances that were left off the canvas.

Many colleagues coaxed and helped us as *Storms from the Sun* took over our lives. For their patience, ideas, comradeship, editorial ideas, and support over the years, we thank Bob Hoffman, Nicky Fox, Mario Acuña, Mauricio Peredo, Jean Desselle, Elaine Lewis, Don Michels, Pål Brekke, Joe Gurman, John Lyon, Michael Wiltberger, and Chuck Goodrich. David Stern, Richard Thompson, and Ed Cliver probably do not realize how much they helped us, but they are all excellent historians of the field who have done a great service to the community by writing and preserving the past. Without James A. Van Allen and his pioneering work there probably wouldn't be a book to write about space weather, so we are particularly grateful for his contribution to this book and to the world. Dan Baker deserves heartfelt thanks for his critical review and thoughtful comments on our manuscript, as does Debra Hudak for a fine job of copy editing. Other colleagues have graciously shared their insights or generously offered their time to review sections of the book, including Joe Allen, George Siscoe, Nancy Crooker, Jeff Hughes, JoAnn Joselyn, Ray Roble, Greg Ginet, Harry van Loon, Gary Heckman, Ernie Hildner, Tom Bogdan, Dave Webb, Tom Holzer, Joe Kunches, George Withbroe, Geoff Reeves, Barbara Poppe, Terry Onsager, Anatta, Joe Hirman,

John Kappenman, Mervyn Freeman, Marsha Korose, David Boteler, Chris Kunstadter, Alan Tribble, Gautam Badwar, Robert Barish, and many others. We also thank Jan Curtis, Dick Hutchinson, and Fred Espenak for use of their photos, and Steele Hill for his artwork, his image magic, and his friendship. We also owe a special debt of gratitude to our favorite intern and library researcher, Theresa Valentine, who probably doesn't realize yet that she would make an excellent reporter if she weren't such a promising young scientist.

Two other mentors and friends contributed in subtle ways to this work. Brother Regis Moccia deserves our gratitude for starting a young writer on his professional journey many years ago, and Steve Maran offered the encouragement—and the occasional friendly chastisement—to get this story out to the public. Ann Merchant, Robin Pinnel, Stephen Mautner, and the rest of the staff of the Joseph Henry Press have made our first foray into book publishing a memorable and enjoyable one. We also owe a special debt of gratitude to our literary agent, Skip Barker, and to our editor, Jeff Robbins, for sticking with us patiently through a long and arduous process, for making the book better, and for believing in us.

Finally, we would like to thank our parents and families for a lifetime of support and a couple of years of extra patience, curiosity, understanding—and babysitting. But most of all, this book would not have been possible without the help of our wives, Florence Langford Carlowicz and Ellen Florian Lopez, who helped us through the joys, the struggle, the excitement, the writer's blocks, and the manic bursts of creativity that brought *Storms* into the world. This book could not have been published without them.

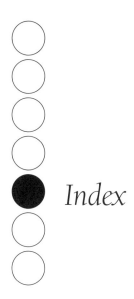

Index

A

Active Cavity Radiometer Irradiance Monitor (ACRIM) experiment, 165
Acuña, Mario, 178
Advanced Composition Explorer (ACE) research satellite, 105, 172, 174-175, 180, 182, 184, 189, 216
Advanced Satellite for Cosmology and Astrophysics, 185-186
Aerospace Corporation, 119
Air Line Pilots Association, 150
Air traffic control, 26
Airlines, commercial flights, 4, 25-26, 86, 91, 110, 113, 149-151, 192-193, 194-195
Alamo, 38
Alfvén, Hannes, 67-68
Allen, Joe, 23, 27, 122-123, 134
American Civil War, 37
American Science and Engineering, 72
American Telephone and Telegraph, 110
Amos (Hebrew prophet), 33-34
Ampère, André-Marie, 63
Anasazi culture, 164
Anaxagoras, 45
Anik E1 and E2 satellites, 23, 121-123, 124, 155

INDEX

Antarctic Circle, 20
Antarctica, 16-19, 44, 138
Antiochos, Spiro, 87
Apollo missions, 137, 139, 140-141, 143-144
Arctic Circle, 20, 138
Aristotle, 37, 40, 45, 46
Armagh Observatory, 169
Astronauts, radiation hazards, 5-6, 20, 86, 88, 91, 137, 139, 140-149, 185
AT&T, 121, 131-133, 138, 179
Auroral electrojet, 16, 96, 101
Auroras
    Aurora borealis (northern lights), ix, 4, 16, 17, 40, 41-42, 56-57, 58, 96-97, 130, 159, 205, 206, 212, 213
    Aurora australis (southern lights), ix, 16-19, 40, 57, 96
    background, 18
    ballad, 61
    cause and formation process, 17-18, 42-44, 49, 55, 56, 83, 85, 88, 89
    and communications disruptions, 57-59, 91
    and compass deviations, 62, 212
    conjugates, 44
    coronas, 109
    displays, ix, 18-19, 40-41, 56, 96, 109, 138
    experimental, 43, 64-65
    and false missile launch indicators, 130
    frequency of displays, 44, 159, 161
    Great Auroras, 44-45, 55-57, 96-97, 107-108, 109, 157
    historical accounts of, 40-41, 42, 56-57
    imaging, 184, 186, 187
    locations of sightings, 4, 16-17, 44-45, 56-57, 65, 96-97, 138, 140, 157, 180, 212
    monitoring, 177, 184, 186
    myths and legends, 40, 41-42
    radiation hazards, 111, 146, 150
    sunspots and, 161, 162
    and weather prediction, 42
    web sites, 205, 206

B

Babcock, Horace, 155-156
Badhwar, Gautam, 5-7, 140, 145, 147, 215
Baker, Dan, 19, 20, 21, 22, 23, 25, 28, 90, 123, 178, 181
Baker, James, 24
Banks, Joseph, 44
Barish, Robert, 150, 151
Barlow, W. H., 57-58
Barnes, Paul, 103
Battle of Hastings, 38
BBC World Service, 113
Biermann, Ludwig, 68
Birkeland, Kristian, 64-65, 66, 67-68, 212, 213
Bogdan, Tom, 85
Bonner Ball neuron detector, 6
Boteler, David, 101, 105, 214
Brekke, Pål, 180, 188, 197
British Antarctic Survey, 17, 177-178
Brooks, John, 7, 107, 108, 109, 110, 111
Butler, John, 169, 170

C

Canadian Broadcast Company, 121
Canadian Geological Survey, 101
Canadian Space Agency, 177
Carrington, Richard, 51-54, 55, 61, 72, 79, 85, 95, 111, 155, 171, 180
Cassini, Giovanni, 161
Catholic Church, 45, 46
CBS News, 11, 25
Celsius, Anders, 43, 62, 63, 212
Central Florida Astronomical Society, 96-97
Chaos theory, 183
Chapman, Sydney, 66-67, 68, 71
Cheyenne Mountain Operations Center, 97
Chinese Television Network, 25
Christofilos, Nicholas, 72-73
Chromosphere, 35
Chronometer, 212

Civilization, celestial events and, 32, 33, 34, 35, 39
Clarke, Arthur C., 75, 117, 193
Climate change
  C^{14} in tree rings, 162-164
  cosmic rays and, 163, 167, 168
  global warming/greenhouse effect, 166, 170, 197
  interplanetary magnetic field and, 167-168
  Little Ice Age, 163, 166, 167
  solar cycles and, 85, 91, 153, 161-164, 197
  solar radiative output and, 164-167
  solar-stratospheric relationship, 165-166, 168-169
  ultraviolet radiation and, 165-166
Cluster II, 89, 177, 216
CNN Airport Network, 25
Cold War, 70, 72-73, 130
Columbus, Christopher, 31-32
Comets, 37-40, 49, 161
Communications disruptions. *See also* Radio waves; Satellites
  airline industry, 110, 113, 193
  backup systems, 28
  broadcast radio and television, 4, 8, 11-12, 13, 18-19, 108, 110, 113-114, 118, 121, 131-133
  CMEs and, 91
  LORAN system, 113
  magnetic storms and, 98-99, 112-113, 131-133
  pager network failure, 11-12, 22-25
  solar flares and, 86
  sunspots and, 85
  telegraph, 54, 57-59
  telephone and teletype, 110, 118, 121, 138, 193
Compasses, 61-62, 97, 212
Concorde, 149
Cook, James, 44
Coronagraphs, 35-36, 71-72, 81
Coronal mass ejections (CMEs), 4, 5
  and auroras, 44-45
  cause and formation process, 80, 86-87, 142, 187
  composition, 90, 127
  coordinated studies of, 178-179, 196
  defined, 13
  discovery, 54, 138
  Earth impacts, 13, 14-15, 16, 17-18, 19, 21, 87-89, 91, 95, 131, 138, 179, 182-183, 185, 187, 193, 196
  force of, 13, 87-89, 91, 185
  geoeffectiveness, 88-89
  halo, 95, 184
  prediction, 188
  rate of occurrence, 87, 94
  size of cloud, 14, 86-87, 179
  solar cycle and, 155, 157
  and solar wind, 8, 14, 36, 75, 87
  time-lapse photos, 71-72
Cosmic rays, x, 69, 111, 127, 142, 148, 149, 150, 163, 167, 168
CS-3B communications satellite, 98
Cyanogen, 38-39

D

Danilov, Valentin, 215
Danish Meteorological Institute, 169
Danish Space Research Institute, 167
Darwin, Charles, 63
de Coulomb, Charles Augustin, 63
De Maculis in Sole Observatis, 46
DeForest, Craig, 80
Descartes, René, 43
Dessler, Alex, 66
Dielectric charging hypothesis, 22-24, 121-124
Dielectric materials, 123-124
DOSMAP dosimetric mapping experiment, 5
Dressler, Alex, 213
Dynamics Explorers, 176

E

Earth Radiation Budget (ERB) experiment, 165
Earthquakes
  celestial events and, 35-36, 179
  monitoring network outage, 132

Eclipses, 31-36, 49, 81, 162, 205, 206
Economic impacts, 98-99, 100, 103-104, 118-119, 122, 128, 132, 193
Eddy, Jack, 160, 161, 162, 163-164, 166
Electric power disruptions and hazards
  causes of, 91, 98, 101-102, 104, 105, 109, 111, 175
  design-basis events, 105
  dielectric charging hypothesis, 22-24
  direct current vs. alternating current, 101-102
  economic impacts, 98-99, 100, 103-104
  electromagnetic pulse and, 73
  forecasting, 104-105, 175, 185
  geology and, 102
  GICs and, 98, 102, 104, 105, 109, 111, 175
  Hydro-Quebec blackout, 99-101
  hypothetical worst-case event, 103-104
  mitigation measures, 4, 104-105, 193
  North American system vulnerability, 101-102, 109-110, 214
  nuclear power plants, 98-99, 175, 194
  power-sharing structure and, 104
  on satellites, 22-24
  for telegraph operators, 59-60, 214
  transformer failure cycle, 102, 138, 140
  vulnerability to, 8, 159, 194
Electrical energy from solar storms, 16, 85, 91
Electromagnetic furnace, 212
Electromagnetic pulse, 73
Electromagnetic waves, x, 63, 79
Electromagnetism, 62-63, 80, 101, 181-182
Eponym Canon, 34
Equator-S satellite, 22
Escouber, Philippe, 89
Eskimos, 42
European Space Agency, 79, 89, 174, 177, 180, 195, 205
European Union, 150
Explorer satellites, 69, 70, 71, 141, 176, 214
Extravehicular activity (EVA), 5, 145-146, 148
Eyde, Sam, 212
Ezekiel (prophet), 41

F

Fabricius, Johannes, 46
Far Ultraviolet Spectroscopic Explorer (FUSE), 126
Faraday, Michael, 63, 101
Fast Auroral Snapshot Explorer (FAST), 180, 216
Federal Aviation Administration, 26, 114, 150
Federal Communications Commission, 27-28
Ferraro, Vincent, 67, 68, 71
Flamsteed, John, 161
Fox Indians of Wisconsin, 42
Franklin, Benjamin, 43
Free University of Berlin, 168
Friis-Christensen, Eigil, 169

G

Galaxy satellites, 12, 20, 22-26, 27, 120, 121
Galilei, Galileo, 45, 46, 47, 153, 161, 171
Galileo Project, 206
Gamma radiation, 77, 86, 144
Gauss, Carl Friedrich, 63
Geiger counters, 69-70
Genesis mission, 181
Geomagnetically induced currents (GICs), 98, 102, 104, 105, 109, 111, 175
Geomagnetism research, 61-67
Geophysics, 62, 63, 67
Geospace, 12
Geotail mission, 177, 216
Gilbert, William, 61-62, 64
Global Positioning System, 8, 98, 114, 125, 128-129, 193

Index

GMS satellites, 22
Goddard Institute for Space Studies, 166
Goddard Space Flight Center, 12, 178, 184
Gold, Thomas, 67
Golightly, Mike, 143, 144, 145, 146, 148
Goodrich, Charles, 182-183
Graham, George, 62, 63, 212
Great Comet of 1861, 37
Gringauz, Konstantin, 71
Gurman, Joe, 190

H

Hairy stars, 37
Hale-Bopp comet, 39, 179
Hale, George Ellery, 63-64
Halley's comet, 37-39, 161
Harriot, Thomas, 46
Harrison, John, 212
Harvard Radio Astronomy Station, 108
Hathaway, David, 157, 158
Heaven's Gate cult, 39
Heckman, Gary, 174-175
Helios, 176
Helms, Susan, 4
Henry, Joseph, 56, 63
Herschel, William, 160
Hey, James Stanley, 115
High-Altitude Observatory (HAO), 71-72, 85, 108, 162, 205
High-energy particles. *See also* Radiation hazards
  detection, 69-70
  dielectric charging hypothesis, 22-24, 121-123
  electron-beam theory, 66
  killer electrons, 12, 23-24, 90, 123, 197
  penetration of spacecraft, 86
  solar proton events, 15, 86, 94, 96, 124-126, 127, 138, 140-149, 159, 184-186
  sources, 15, 85, 86-90
  in Van Allen radiation belts, 19, 89-90, 91

High-Energy Solar Spectroscopic Imager (HESSI), 181
Hildner, Ernie, 168, 173, 175
Hiortier, Olof, 62, 212
Historical accounts
  of auroras, 40-41, 42, 162, 163
  of comets, 37-39, 161
  of eclipses, 33, 34, 162
  of sunspots, 45-46, 161-162, 163
HMS Endeavour, 44
Hodgson, Richard, 52-53
Hoffman, Robert, 179
Hoffmeister, Cuno, 68
Hubble Space Telescope, 125-126, 128
Hudson, Donald, 150, 151
Hughes Space and Communications, 24, 26, 121
Hyakutake comet, 39

I

Igneous rock geologies, 102
Imager for Magnetopause to Aurora Global Exploration (IMAGE), 180, 184
Institute of Space and Astronautical Science (ISAS), 177
Intelsat-K satellite, 123
International Geophysical Year (1957-1958), x, 69, 108, 111, 214
International Solar-Terrestrial Physics program, 19, 176-181, 184, 187, 216
International Space Station *Alpha*, 1, 4, 5, 6, 20, 145-148, 185, 195
International Sun-Earth Explorers, 176
Internet, 187, 193-194
Interplanetary Monitoring Platforms, 176, 216
Inuit, 41-42
Ion traps, 71
Ionosphere, x, 16, 18-19, 20, 68, 86, 89, 96, 98, 142
Ionospheric scintillation, 112-114, 129, 192
Iridium satellites, 22

225

J

Jet Propulsion Laboratory, 181
John (evangelist), 34
John of Worcester, 45
Johns Hopkins University, 126, 181
Johnson Space Center, 5, 140, 143, 145, 173
Joselyn, JoAnn, 93-94, 159, 196

K

Kappenman, John, 101, 104-105
Kelvin, William Thomson (Lord), 64
Kennard, William, 27-28
Kepler, Johann, 39
Kew Observatory, 54
Khrushchev, Nikita, 71
Killer electrons, 12, 23-24, 90, 123, 197
Kirkwood, Daniel, 56-57
Kongespeilet, 42
Korolev, Sergei, 70, 71
Kunstadter, Christopher, 118, 119

L

Labitzke, Karin, 168
Lanzerotti, Louis, 133
Large-Angle Spectrometric Coronagraph, 15-16, 125
Larson, Michelle Beauvais, 78
Lassen, Knud, 169
Little Ice Age, 163, 166, 167
Living with a Star program, 195-196
Lockheed Martin, 132, 146
Lockwood, Mike, 167-168
Lodestone, 61
Long-Range Navigation (LORAN) system, 113
Loomis, Elias, 56, 58, 61, 63, 72, 111, 180, 212
Loral Space & Communications, 132
Los Alamos National Laboratory, 20-21, 90, 177
Louis I, 35
Lucent Technologies, 133
Ludwig, George, 69
Lunar base, 148, 192
Lunik spacecraft, 71
Lyot, Bernard-Ferdinand, 35-36

M

MacQueen, Bob, 71-72
Magnetic field lines, 14, 19, 43-44, 65, 81, 87, 89
Magnetic fields
 Earth's, *see* Geomagnetism; Magnetosphere
 interaction, 87, 89, 146
 interplanetary, 83, 167-168
 in solar wind, 68-69, 82, 83
 Sun's, 3, 82, 84, 87, 89, 153, 154-155, 156, 168
Magnetic observatories, 51, 52, 63
Magnetic reconnection, 87, 89, 183, 187
Magnetic storms, xi. *See also* Communications disruptions; Satellites
 1847, 57-58
 1859, 54-57, 58-60, 111, 212
 1909, 44
 1958, 7, 107-112
 1989, 44, 95-96, 97-101, 102, 105, 113, 122, 127, 145, 149, 213
 1991, 127
 1997, 131-133
 1998, 12-16, 18-19, 21, 22, 24, 28, 118-119
 2000, 81, 118-119, 147, 157, 175, 184-186
 2001, 1, 2-5
 alerts, 97, 120
 cause and formation process, 89
 CMEs and, 88-89
 equatorial, 66, 67
 first direct observation, 62
 hypothetical worst-case, 103-104
 solar cycle and, 138, 154, 155
 solar flares and, 85
 solar wind and, 83, 174-175
 sunspots and, 48, 55-56, 59-60, 64, 83-84, 85

MagnetoHydroDynamics equations, 182
Magnetosphere, Earth
  and auroras, 17-18, 43-44
  "cavity," 13, 16, 67, 71, 83
  CMEs and, 88-89, 179, 182-183, 185
  compression, 16, 96, 97, 131, 146, 179, 185
  electrical energy, 83
  killer electrons in, 24
  MHS simulations, 182-184, 196
  monitoring, 89, 177
  nuclear weapons tests in, 72-73
  plasma particle detection in, 71
  plasma sheet, 88
  radiation hazards in, 145, 148
  solar cycle and, 156
  solar flares and disruption of, 54-55, 67, 138
  solar wind penetration, 83, 86
  South Atlantic Anomaly, 20, 125-126, 142
  Van Allen radiation belts and, 19-20, 88, 179
  Web sites, 203-204
Maran, Steve, 213
Marecs-A navigation satellite, 127
Mariner II, 71
Mars exploration, 148, 149, 192
Mars Odyssey spacecraft, 4, 215
Mars Radiation Environment Experiment, 215
Marshall Space Flight Center, 157
Mason, Helen, 82
Maunder, E. W., 64, 71, 72, 160-161, 162-163
Max Planck Institute, 177
Maxwell, James Clerk, 63, 64, 182
McIlwain, Carl, 69-70
Medicine Wheels, 33
Medieval Climatic Optimum, 164
Mercury (planet), 45, 47, 48
Michels, Don, 178-179
Microwaves, 85
Mid-Atlantic Area Council power pools, 100
Milton, John, 34-35
Minnesota Power and Electric, 102

Mir space station, 7, 127-128, 144, 145, 185
Montana State University, 78
Moos, Warren, 126
Motorola, 22
Muir, John, 93
Myths and legends
  of auroras, 40, 41-42
  of comets, 37-38, 49
  of eclipses, 32, 34-35, 49

N

National Aeronautics and Space Administration (NASA). *See also specific missions and spacecraft*
  ESA joint missions, 79
  flight rules, 147-148
  launch/mission delays, 4, 140
  monitoring projects, 189
  radiation risk study, 146-147
  Solar Cycle 23 Project, 156-157
  Space Radiation Analysis Group, 5, 143, 148, 173
  spacecraft design, 195
  web sites, 205
National Center for Atmospheric Research, 168
National Geophysical Data Center, 122
National Oceanic and Atmospheric Administration, 24, 94, 98, 119, 120, 122, 124, 127, 137, 147, 148, 156-157, 159, 168, 172, 173, 177, 181, 195, 205
National Public Radio (NPR), 11, 25
National Research Council, 146-147
National Science Foundation, 178, 189, 195
National Security Space Architecture, 129
National Solar Observatory, 205
National Weather Service, 173
Native American Indian tribes, 41-42
Neumann, Peter, 28-29
Newton, Isaac, 63
Nimbus-7 satellite, 165

INDEX

Nobel Prizes, 66, 68
Norsk Hydro, 212
North American Electric Reliability Council, 103
Northeast Power Coordinating Council, 100
Nuclear fusion, 77
Nuclear weapons tests, 72-73

O

Oak Ridge National Laboratory, 103
Occupational Safety and Health Administration, 143
Oersted, Hans Christian, 63
Ogilvie, Keith, 179
Oil prospecting operations, 97
Onsager, Terry, 183, 188
Operation Argus, 72-73
Orbiting Solar Observatory 7, 71
Orr, Eric, 172-173
Ozone layers, 166-167

P

Pager network failure, 11-12, 22-25
Paley, Matthew, 17-19, 21
PanAmSat Corporation, 12, 24, 26-27, 211
Paré, Ambroise, 37
Parker, Eugene, 68-69, 71, 72, 161
Path of totality, 35
Peloponnesian War, 34
Penumbrae, 3, 45
Phantom Torso experiment, 6-7
Phlegon, 34
Physio-Meteorological Observatory (Havana, Cuba), 57
Picard, Jean, 161
Plague of 1665, 37
Planetary Society, 195, 206
Plasma
  auroral flows, 177
  composition, 49, 67
  coronal flows, 82, 86
  energy transport, 78, 80
  formation, 76
  ionospheric, 112-113

particle detectors, 71
pervasiveness, 212
physics, 67-68
properties, 35, 89, 112-113
surface charging of spacecraft, 126-127
in Van Allen radiation belts, 19, 89, 188
Photospheric granules, 3
Plutarch, 37
Poey, Andreas, 57
Poincaré, Henri, 64
Polar Orbiting Environmental Satellites, 21, 22, 177, 184, 186, 216
Pompeii, 37
Pope Callixtus III, 38
Prescott, George B., 58-59
Printy, Tom, 96-97
Prophecies and predictions of celestial events, 31-32, 33-34, 38
Ptolemy, Claudius, 37
Public Service Electric and Gas Co., 98-99

R

Radar, 114-115
Radiation hazards
  to airline passengers and crew, 4, 86, 149-151, 194-195, 206
  to astronauts, 5-6, 7, 8, 20, 86, 137, 139, 140-149, 185, 192, 197
  in auroral zones, 111, 146, 150
  cancer risk, 143, 144-145, 146-147, 149, 150, 151
  from cosmic rays, 141, 142, 148, 149, 150
  detection/detectors, 69-70
  dose measures, 215
  exposure limits, 141, 143, 146-147, 150-151
  genetic mutation, 143, 149
  high-energy particles (ionizing), 12, 15, 23-24, 86, 142-143
  in ionosphere, 142
  lethal dose, 143, 145
  long-term, low-level doses, 148-149

in magnetosphere, 145, 148
  modeling, 144-145
  monitoring, 5, 6-7
  nuclear weapons tests and, 73
  to pregnant women, 150
  radiation sickness, 142-143
  to satellites, 119, 124, 125-126, 127-128, 129
  shielding and, 20, 22, 23, 28, 119, 126, 129, 130-131, 134-135, 140, 145, 146, 147, 214-215
  from solar flares, 86, 127, 137, 139, 140-141, 145, 150
  from solar proton events, 86, 124, 127, 139, 140-149
  from Van Allen radiation belts, 20-21, 73, 126, 142, 146
  warning of, 142, 148
Radio Free Europe, 113
Radio waves
  applications, 112, 187
  communications distortions and blackouts, 4, 8, 11-12, 18-19, 85, 108, 110, 113-114, 130, 138, 184, 192-193
  discovery, 112
  ionospheric scintillation, 112-114
  principles of communication, 112
  projected interruptions, 129, 192-193
  solar emissions, x, 13, 85, 108-109, 114-115, 138, 187
Ramey Solar Observatory, 93
Ray, Ernie, 70
RCA Communications, 110
Reeves, Geoff, 20-22, 23, 24, 90
Reuters news service, 25
Ring current, 16, 68
Roberts, Walter Orr, 108
Rostoker, Gordon, 24
Royal Astronomical Society, 64, 160
Royal Greenwich Observatory, 160
Royal Society of London, 64
Royce, Frederick, 59
Russell, Chris, 22
Russian Space Research Institute, 177
Rust, Dave, 72
Rutherford Appleton Laboratory, 167
Rutja, 41

S

Sabine, Edward, 48, 51, 55, 56, 63
Sacramento Peak Solar Observatory, 108
Saint George, 41
Sandia National Laboratory, 214-215
Saros cycles, 32, 33
Satellites. *See also specific satellites and missions*
  applications, 117
  atmospheric drag, 96, 97, 98, 128, 186
  backup systems, 27, 121, 135-136, 184
  CMEs and, 88, 96
  and commerce, 118
  communication problems of, 91, 98, 114, 121, 127, 128, 159, 184
  communications-related, 8, 11-12, 20, 22, 23-25, 26-27, 98, 117-118, 121-123, 131-133, 157, 179
  data recorders, 69, 120
  deep dielectric charging, 22-24, 121-124
  design and construction shortcomings, 127, 129, 130-131, 194
  dielectric materials, 123-124
  economic investment, 117
  electrostatic discharge, 126-127
  environment-related losses, 119-120, 128-133
  failure rates, 27-28, 118, 133, 175
  forecasting and, 175
  in geostationary orbit, 14, 94, 98, 115, 123, 128-129
  geosynchronous orbit, 16, 20, 96, 185
  GPS, 8, 98, 114, 125, 127, 128-129, 191, 193
  hardened/shielded, 20, 22, 23, 28, 119, 126, 129, 130-131, 134-135, 194, 211
  HS 601 model, 26
  insurance claims, 118-119, 132
  ionospheric scintillation and, 114

INDEX

killer electrons, 12, 23-24, 90, 123
low-Earth orbit, 128, 129, 196
military, 4, 117, 128-131
miniature, 196
mitigation measures, 22, 119
number in orbit, 28, 117, 118
orbital displacement, 96, 97, 98, 128, 131
outages and failures, 4, 8, 11-12, 20, 22-27, 73, 86, 91, 98, 119, 121-129, 157, 159, 179, 184-186, 193, 206
radiation damage, 119, 126
reporting of anomalies, 133-134
scientific, 22, 117, 125-126, 129, 165, 184-186
single-event upsets, 124-126
solar cycle and threat to, 28, 155
solar flares and, 86, 127
solar panel failures, 127-128, 185-186
Starfish nuclear weapons test and, 73
surface charging, 126-127
tin whiskers problem, 26-27, 120
Van Allen radiation belts and, 20, 22-23, 73, 90, 124, 127, 129, 135
weather-tracking, 14, 26, 94, 98, 123, 127-128, 129, 165
Saturn, 161
Scheiner, Christopher, 46, 161
Schou, Jesper, 79
Schwabe, Heinrich, 47-48, 51, 154
Seneca, 40
Service, Robert W., 61
Shindell, Drew, 166-167
Single-event upsets, 124-126
Siscoe, George, 147-148
Skylab space station, 71, 72, 128, 144
Skynet Satellite Services, 132
Smithsonian Institution, 56
Solar and Heliospheric Observatory (SOHO), 13, 15-16, 79, 125, 172, 174, 177, 180, 184, 188, 189, 190, 216
Solar, Anomalous, and Magnetospheric Particle Explorer (SAMPEX), 19, 21, 216

Solar cycles
11-year, 79-80, 153, 154, 160, 163, 167, 169
22-year, 154
and auroras, 44, 162
and climate change, 85, 91, 153, 161-167
and CMEs, 155, 157
current cycle 23, 28, 156, 157, 158, 187
discovery, 154
and magnetic storms, 138, 154, 155, 157
Maunder minimum, 160-161, 162-163, 164, 167, 169
maximum, 1, 15, 20, 28, 94, 98, 128, 129, 131, 134, 154, 155, 157, 158, 159, 167, 187, 216
medieval maximum, 164
minimum, 28, 128, 134, 137, 142, 154, 156, 160, 167
NASA/NOAA project, 156-157, 159, 165
and ozone layers (stratospheric), 166
and particle emissions, 153
and polar shift, 154-155
research satellite, 94, 97
and solar flares, 155, 157, 213, 216
and solar irradiance, 164-166, 168-169
and solar wind, 155, 162, 167
and space weather predictions, 156-157, 159
and spacecraft/astronaut risk, 28, 124, 145-146
Spörer minimum, 164
and stratospheric oscillation, 168-169
and sunspot patterns, 48, 51, 55, 79-80, 153, 154-156, 157-158, 162, 216
and terrestrial weather, 168-170
transformer failure cycle and, 102
ultraviolet radiation, 165-166
Solar dynamo, 79-80
Solar eclipses, 32-36
Solar flares
and auroras, 44-45, 55, 85

cause, 13, 80, 81, 85-86, 137-138
discovery, 51-54, 61, 85
emissions, 13, 85, 94, 138
force, 13, 54-55, 85, 94
and magnetic storms, 55, 64, 85
M-class, 94
radiation hazards, 86, 127, 137, 139, 140-141, 145, 150
rate of occurrence, 85
seahorse, 138-140
size, 53, 94, 111
solar cycle and, 155, 157, 213, 216
and solar wind, 8, 86, 111
and space-based activities, 86
spectroheliographs, 63-64
sunspots and, 85, 94, 142
temperature, 94
"white-light," 2-3, 4-5, 52-54, 95, 108, 127
X-class, 93, 94, 138-140, 145, 192, 213
X-ray intensity, 3, 54, 138, 213
Solar Maximum Mission (SMM) satellite, 94, 97, 165, 176
Solar proton events, 15, 86, 94, 96, 124-126, 127, 138, 140-149, 159, 184-186, 192-193
Solar Proton Modeling Experiment, 141
Solar sail, 195, 196, 206
Solar wind
 and auroras, 43-44
 CMEs and, 14
 comets' tails and, 39-40, 68
 composition, 8, 66, 67, 76, 83, 89-90, 90
 cycle, 71, 72
 defined, x
 density, 66, 83
 detectors, 71
 magnetic field in, 68-69, 82, 83, 167
 magnetosphere penetration, 83, 90
 mass, 82-83
 monitoring and modeling, 104-105, 172, 174-175, 177, 181, 184, 188, 196
 origin, 36, 72, 81, 82
 speed, 16, 54, 64, 82, 86, 120, 196
 structures, 162
 and Van Allen radiation belts, 21
Solar X-ray Imager (SXI), 181
South Atlantic Anomaly, 20, 125-126, 142
Southwest Research Institute, 80
Space Environment Center (SEC), 94, 119, 120, 122, 124, 137, 148, 159, 168, 172, 173, 177, 183, 184, 188, 189, 195, 196, 205
Space lightning, 126-127
Space physics, 63, 129-130
Space shuttle
 Atlantic, 145
 Endeavor, 5, 6
 mission STS-100, 5
 radiation hazards, 20, 148
Space telescopes, 81, 13, 15-16, 79
Space tourism, 192, 195
Space weather. *See also* Magnetic storms; Space weather forecasting
 causes, x, 36, 55, 80, 83-84, 85
 engineering responses, xi
 first widely observed instance, 54-57, 58-59, 63
 hypothetical worst-case event, 103-104
 ignorance of, 129-130, 135, 179-180
 information deficiencies, 135
 insurance claims, 118-119
 most significant, 101
 practical applications, x
 reporting of failures due to, 120, 133-134
 and terrestrial weather, 168-169, 181, 214
 theoretical basis, 67-68
 visible manifestations, ix, 3-4, 16-19, 44-45, 49
 vulnerability of society to, x-xi, 4, 7, 8-9, 20, 25-29, 49, 59, 75, 98-101, 159, 191-195
 Web sites, 203-205
Space Weather Architecture Study, 129, 134

Space weather forecasting
  accuracy, 174-175, 175, 196
  alerts, watches, and warnings, 142, 147, 173, 175, 185, 187
  for electric power companies, 104-105, 175, 185
  engineering interface with, xi
  funding, 188-189, 196
  International Solar-Terrestrial Physics program and, 19, 176-181
  maps, 21-22, 181
  MHD simulations and, 181-184
  planetary K index, 173
  for satellite operators, 175
  Solar Cycle 23 Project, 156-157
  solar proton events, 142, 159
  solar wind monitoring and, 105, 174-175, 182-184
  Space Weather Operations Center, 171-172
  tools for, 143, 147, 159, 172, 174-175, 176-181, 187-190, 196
  vulnerability of equipment, 184-186
Spacecraft design, 127, 129, 130-131, 194, 195-196. *See also* Satellites; *individual spacecraft*
Special world interval advisory, 108
Spectroheliograph, 63-64
Speich, Dave, 119, 122-123
Spörer, Gustav, 160, 162
Springsteen, Bruce, 171
Sputnik satellites, 69, 70-71, 176
SRI International, 28
Stanford University, 79
Starfish High Altitude Nuclear Test program, 73
Stewart, Balfour, 51, 55-56, 59-60, 63, 72
Stoermer, Carl, 65-66
Stonehenge, 33
Sullivan, Walter, 109, 111-112
Sun
  atmosphere, 8, 35-36, 46, 71, 76, 187
  changes in, 7-8, 167, 187; *see also* Solar cycle
  chromosphere, 80, 85
  composition, 77, 78
  convection zone, 78-79, 80
  core, 79
  corona, 35-36, 68, 71-72, 80-82, 86, 120, 155, 156, 162, 196
  distance to Earth, ix, 77
  energy transport in, 78-79, 167
  far-side imaging and modeling, 187
  fusion process, 77
  magnetic fields, 3, 82, 84, 87, 89, 153, 154-155, 156, 168
  nonthermal emissions, x, 4, 77, 153
  photosphere, 35, 78, 80, 82
  radiation zone, 78, 79, 80
  radiative output, ix, x, 165-167
  radio noise from, 108
  rotation rates, 51, 64, 79, 155-156
  safe viewing, 2, 52
  seasons, *see* Solar cycle
  seismic waves, 85-86, 108
  size and age, 76-77, 213
  temperatures, ix, 78, 82, 84, 196
  toroidal fields, 79-80, 81-82, 156
Sunspots
  active region 5395, 93-94, 95
  active region 5935, 95, 97
  active region 8210, 14-16
  active region 9077, 184
  active region 9393, 2-3, 4
  active region 9415, 3, 4-5
  active region 9433, 3-4
  active region ABOO, 108
  and auroras, 161, 162
  cause, 49, 80, 84, 85
  and climate change, 85
  and CMEs, 142
  cycle, 48, 51, 55, 154-156, 169, 216
  discovery and early studies, 45-48, 79, 84
  emissions, 85
  location, 51
  and magnetic storms, 48, 55-56, 59-60, 64, 83-84, 85
  prolonged minimum, 160-161, 162-163, 164, 167
  properties, 83-84
  and solar flares, 85, 94, 142
  structures, 3, 45

and temperatures on Earth, 169
time-series images, 95
Svensmark, Henrik, 167, 170
Swift-Tuttle comet of 1862, 37
Sydkraft, 98

T

Teal Group, 118
Telesat Canada, 121, 122
Telstar satellites, 121, 131-133, 155, 179
Terra Earth Observing Spacecraft, 126
Terrella, 61, 64-65
Thales, 34
Theophrastus, 45
Thermal radiation, ix, x
Thermosphere-Ionosphere-Mesosphere Energetics and Dynamics mission, 181
Thompson, Barbara, 12-13, 21
Thor, 41
Thucydides, 34
Tiberius, 40-41
Tinsley, Brian, 167
Tito, Dennis, 195
Tousey, Richard, 71
Transition Region and Coronal Explorer (TRACE) spacecraft, 81, 174, 180, 184
Treaty of Verdun, 35
Tribble, Alan, 119, 135-136
Triewald, Marten, 43
Tschan, Chris, 129, 130, 131, 134
Twain, Mark, 38

U

Ultraviolet radiation, 83, 165-166
Umbrae, 3, 45
University of California at Los Angeles, 22
University of Chicago, 68, 161
University of Colorado, Laboratory for Atmospheric and Space Physics, 21, 90, 123, 178
University of Iowa, 19, 69, 70
University of Maryland, 182
University of Minnesota, 111
University of Texas at Dallas, 167
Upper Atmospheric Research Satellite (UARS), 165, 166
U.S. Air Force, 97, 122, 129, 171-172, 177, 205
U.S. Aviation Underwriters, 118
U.S. Coast Guard LORAN system, 113
U.S. Department of Defense, 114, 128-131
U.S. Geostationary Operational Environment Satellites (GOES), 14, 94, 98, 123, 127, 181, 184
U.S. National Bureau of Standards, x
U.S. Naval Research Laboratory, 71, 87, 178-179
U.S. Nuclear Regulatory Commission, 175
Usachev, Yury, 4

V

Van Allen, James, 19, 20, 69-70, 71, 214
Van Allen radiation belts
composition, 19
discovery, 19, 69-70
dose modeling, 135, 196
intensity changes, 20-21, 89-90
new belts, 21, 73
nuclear weapons tests and, 72-73
particle accumulation and acceleration in, 19, 89-90, 91, 124-126, 131, 138, 179, 196, 197
radiation hazards from, 20-21, 69-70, 73, 89, 127, 135, 142, 146
solar wind and, 21, 188
and spacecraft, 20, 22-23, 91, 127
structure, 19-20, 70, 90
weather maps of, 21-22
Van Dyke, James, 103
van Loon, Harry, 168, 170
Venus, 45, 71
Verne, Jules, 193
Vernov, Sergei, 70-71
Vesuvius eruption, 37
Vikings, 42

INDEX

Voice of America, 113
Volcanoes, and auroras, 43
von Humboldt, Alexander, 48, 51, 62, 63
Voss, James, 4
Voyager 2 spacecraft, 192
Vulcan, 45, 47, 48

W

Watson-Watt, Robert, 114
Weapon systems control, 114
Weber, Wilhelm, 63
Weyland, Mark, 146
William the Conqueror, 38
Wind spacecraft, 16, 177, 182, 216
Winkler, John, 111
Withbroe, George, 187
Wood, O.S., 58
World Data Center on Solar Activity, 108
World War II, 114-115
World Warning Agency, 109

X

X-ray telescope, 72, 181
X rays
  auroras and, 111
  and communications equipment, 91
  dose to humans, 5-6, 144
  intensity of solar flares, 3, 54, 85, 86, 138
  speed, 184
  sunspot emissions, 85

Y

Yale University, 56
Yohkoh satellite, 174, 184, 204

Z

Zurich Observatory, 46-47